NATEF Standards

Electrical and Electronic Systems (A6)

Fourth Edition

Jack Erjavec
Ken Pickerill

CENGAGE
Learning·

Australia • Brazil • Japan • Korea • Mexico • Singapore • Spain • United Kingdom • United States

NATEF Standards Job Sheets
Electrical and Electronic Systems (A6)
Fourth Edition
Jack Erjavec & Ken Pickerill

VP, General Manager, Skills and Planning:
Dawn Gerrain

Director, Development, Global Product
Management, Skills: Marah Bellegarde

Product Manager: Erin Brennan

Senior Product Development Manager:
Larry Main

Senior Content Developer: Meaghan Tomaso

Product Assistant: Maria Garguilo

Marketing Manager: Linda Kuper

Market Development Manager: Jonathon
Sheehan

Senior Production Director: Wendy Troeger

Production Manager: Mark Bernard

Content Project Management: S4Carlisle

Art Direction: S4Carlisle

Media Developer: Debbie Bordeaux

Cover image(s): © Snaprender/Dreamstime.com
© Shutterstock.com/gameanna
© i3alda/www.fotosearch.com
© IStockPhoto.com/tarras79
© IStockPhoto.com/tarras79
© Laurie Entringer
© IStockPhoto.com/Zakai
© stoyanh/www.fotosearch.com

For product information and technology assistance, contact us at
Cengage Learning Customer & Sales Support, 1-800-354-9706
For permission to use material from this text or product,
submit all requests online at **www.cengage.com/permissions.**
Further permissions questions can be e-mailed to
permissionrequest@cengage.com

Library of Congress Control Number: 2014931166

ISBN-13: 978-1-1116-4702-5
ISBN-10: 1-111-64702-X

Cengage Learning
200 First Stamford Place, 4th Floor
Stamford, CT 06902
USA

Cengage Learning is a leading provider of customized learning solutions with office locations around the globe, including Singapore, the United Kingdom, Australia, Mexico, Brazil, and Japan. Locate your local office at:
www.cengage.com/global

Cengage Learning products are represented in Canada by Nelson Education, Ltd.

To learn more about Cengage Learning, visit **www.cengage.com**
Purchase any of our products at your local college store or at our preferred online store **www.cengagebrain.com**

Notice to the Reader
Publisher does not warrant or guarantee any of the products described herein or perform any independent analysis in connection with any of the product information contained herein. Publisher does not assume, and expressly disclaims, any obligation to obtain and include information other than that provided to it by the manufacturer. The reader is expressly warned to consider and adopt all safety precautions that might be indicated by the activities described herein and to avoid all potential hazards. By following the instructions contained herein, the reader willingly assumes all risks in connection with such instructions. The publisher makes no representations or warranties of any kind, including but not limited to, the warranties of fitness for particular purpose or merchantability, nor are any such representations implied with respect to the material set forth herein, and the publisher takes no responsibility with respect to such material. The publisher shall not be liable for any special, consequential, or exemplary damages resulting, in whole or part, from the readers' use of, or reliance upon, this material.

Printed in the United States of America

1 2 3 4 5 6 7 18 17 16 15 14

CONTENTS

NATEF TASK LIST FOR ELECTRICAL AND ELECTRONIC SYSTEMS

Required Supplemental Tasks (RST)

Shop and Personal Safety

1. Identify general shop safety rules and procedures.

2. Utilize safe procedures for handling of tools and equipment.

3. Identify and use proper placement of floor jacks and jack stands.

4. Identify and use proper procedures for safe lift operation.

5. Utilize proper ventilation procedures for working within the lab/shop area.

6. Identify marked safety areas.

7. Identify the location and the types of fire extinguishers and other fire safety equipment; demonstrate knowledge of the procedures for using fire extinguishers and other fire safety equipment.

8. Identify the location and use of eyewash stations.

9. Identify the location of the posted evacuation routes.

10. Comply with the required use of safety glasses, ear protection, gloves, and shoes during lab/shop activities.

11. Identify and wear appropriate clothing for lab/shop activities.

12. Secure hair and jewelry for lab/shop activities.

13. Demonstrate awareness of the safety aspects of supplemental restraint systems (SRS), electronic brake control systems, and hybrid vehicle high-voltage circuits.

14. Demonstrate awareness of the safety aspects of high-voltage circuits (such as high intensity discharge (HID) lamps, ignition systems, injection systems, etc.).

15. Locate and demonstrate knowledge of material safety data sheets (MSDS).

Tools and Equipment

1. Identify tools and their usage in automotive applications.

2. Identify standard and metric designation.

3. Demonstrate safe handling and use of appropriate tools.

4. Demonstrate proper cleaning, storage, and maintenance of tools and equipment.

5. Demonstrate proper use of precision measuring tools (i.e., micrometer, dial-indicator, dial-caliper).

Preparing Vehicle for Service

1. Identify information needed and the service requested on a repair order.

2. Identify purpose and demonstrate proper use of fender covers, mats.

3. Demonstrate use of the three C's (concern, cause, and correction).

4. Review vehicle service history.

5. Complete work order to include customer information, vehicle identifying information, customer concern, related service history, cause, and correction.

Preparing Vehicle for Customer

1. Ensure vehicle is prepared to return to customer per school/company policy (floor mats, steering wheel cover, etc.).

Maintenance and Light Repair (MLR) Task List

A. General

A.1.	Research applicable vehicle and service information, vehicle service history, service precautions, and technical service bulletins.	Priority Rating 1
A.2.	Demonstrate knowledge of electrical/electronic series, parallel, and series-parallel circuits using principles of electricity (Ohm's law).	Priority Rating 1
A.3.	Use wiring diagrams to trace electrical/electronic circuits.	Priority Rating 1
A.4.	Demonstrate proper use of a digital multimeter (DMM) when measuring source voltage, voltage drop (including grounds), current flow, and resistance.	Priority Rating 1
A.5.	Demonstrate knowledge of the causes and effects from shorts, grounds, opens, and resistance problems in electrical/electronic circuits.	Priority Rating 2
A.6.	Check operation of electrical circuits with a test light.	Priority Rating 2
A.7.	Check operation of electrical circuits with fused jumper wires.	Priority Rating 2
A.8.	Measure key-off battery drain (parasitic draw).	Priority Rating 1
A.9.	Inspect and test fusible links, circuit breakers, and fuses; determine necessary action.	Priority Rating 1
A.10.	Perform solder repair of electrical wiring.	Priority Rating 1
A.11.	Replace electrical connectors and terminal ends.	Priority Rating 1

B. Battery Service

B.1.	Perform battery state-of-charge test; determine necessary action.	Priority Rating 1
B.2.	Confirm proper battery capacity for vehicle application; perform battery capacity test; determine necessary action.	Priority Rating 1
B.3.	Maintain or restore electronic memory functions.	Priority Rating 1
B.4.	Inspect and clean battery; fill battery cells; check battery cables, connectors, clamps, and hold-downs.	Priority Rating 1
B.5.	Perform slow/fast battery charge according to manufacturer's recommendations.	Priority Rating 1
B.6.	Jump-start vehicle using jumper cables and a booster battery or an auxiliary power supply.	Priority Rating 1
B.7.	Identify high-voltage circuits of electric or hybrid electric vehicle and related safety precautions.	Priority Rating 3

B.8. Identify electronic modules, security systems, radios, and other accessories that require reinitialization or code entry after reconnecting vehicle battery.   Priority Rating 1

B.9. Identify hybrid vehicle auxiliary (12v) battery service, repair, and test procedures.   Priority Rating 3

C. Starting System
C.1. Perform starter current draw test; determine necessary action.   Priority Rating 1
C.2. Perform starter circuit voltage drop tests; determine necessary action.   Priority Rating 1
C.3. Inspect and test starter relays and solenoids; determine necessary action.   Priority Rating 2
C.4. Remove and install starter in a vehicle.   Priority Rating 1
C.5. Inspect and test switches, connectors, and wires of starter control circuits; determine necessary action.   Priority Rating 2

D. Charging System
D.1. Perform charging system output test; determine necessary action.   Priority Rating 1
D.2. Inspect, adjust, or replace generator (alternator) drive belts; check pulleys and tensioners for wear; check pulley and belt alignment.   Priority Rating 1
D.3. Remove, inspect, and reinstall generator (alternator).   Priority Rating 2
D.4. Perform charging circuit voltage drop tests; determine necessary action.   Priority Rating 1

E. Lighting Systems
E.1. Inspect interior and exterior lamps and sockets including headlights and auxiliary lights (fog lights/driving lights); replace as needed.   Priority Rating 1
E.2. Aim headlights.   Priority Rating 2
E.3. Identify system voltage and safety precautions associated with high-intensity discharge headlights.   Priority Rating 2

F. Accessories
F.1. Disable and enable airbag system for vehicle service; verify indicator lamp operation.   Priority Rating 1
F.2. Remove and reinstall door panel.   Priority Rating 1
F.3. Describe the operation of keyless entry/remote-start systems.   Priority Rating 3
F.4. Verify operation of instrument panel gauges and warning/indicator lights; reset maintenance indicators.   Priority Rating 1
F.5. Verify windshield wiper and washer operation; replace wiper blades.   Priority Rating 1

Automobile Service Technology (AST) Task List

A. General Electrical System Diagnosis
A.1. Research applicable vehicle and service information, vehicle service history, service precautions, and technical service bulletins.   Priority Rating 1
A.2. Demonstrate knowledge of electrical/electronic series, parallel, and series-parallel circuits using principles of electricity (Ohm's law).   Priority Rating 1
A.3. Demonstrate proper use of a digital multimeter (DMM) when measuring source voltage, voltage drop (including grounds), current flow, and resistance.   Priority Rating 1
A.4. Demonstrate knowledge of the causes and effects from shorts, grounds, opens, and resistance problems in electrical/electronic circuits.   Priority Rating 1
A.5. Check operation of electrical circuits with a test light.   Priority Rating 1
A.6. Check operation of electrical circuits with fused jumper wires.   Priority Rating 1
A.7. Use wiring diagrams during the diagnosis (troubleshooting) of electrical/electronic circuit problems.   Priority Rating 1

E.3.	Aim headlights.	Priority Rating 2
E.4.	Identify system voltage and safety precautions associated with high intensity discharge headlights.	Priority Rating 2

F. Gauges, Warning Devices, and Driver Information Systems Diagnosis and Repair

F.1.	Inspect and test gauges and gauge sending units for causes of abnormal gauge readings; determine necessary action.	Priority Rating 2
F.2.	Diagnose (troubleshoot) the causes of incorrect operation of warning devices and other driver information systems; determine necessary action.	Priority Rating 2

G. Horn and Wiper/Washer Diagnosis and Repair

G.1.	Diagnose (troubleshoot) causes of incorrect horn operation; perform necessary action.	Priority Rating 1
G.2.	Diagnose (troubleshoot) causes of incorrect wiper operation; diagnose wiper speed control and park problems; perform necessary action.	Priority Rating 2
G.3.	Diagnose (troubleshoot) windshield washer problems; perform necessary action.	Priority Rating 2

H. Accessories Diagnosis and Repair

H.1.	Diagnose (troubleshoot) incorrect operation of motor-driven accessory circuits; determine necessary action.	Priority Rating 2
H.2.	Diagnose (troubleshoot) incorrect electric lock operation (including remote keyless entry); determine necessary action.	Priority Rating 2
H.3.	Diagnose (troubleshoot) incorrect operation of cruise control systems; determine necessary action.	Priority Rating 3
H.4.	Diagnose (troubleshoot) supplemental restraint system (SRS) problems; determine necessary action.	Priority Rating 2
H.5.	Disable and enable airbag system for vehicle service; verify indicator lamp operation.	Priority Rating 1
H.6.	Remove and reinstall door panel.	Priority Rating 1
H.7.	Check for module communication errors (including CAN/BUS systems) using a scan tool.	Priority Rating 2
H.8.	Describe the operation of keyless entry/remote-start systems.	Priority Rating 3
H.9.	Verify operation of instrument panel gauges and warning/indicator lights; reset maintenance indicators.	Priority Rating 1
H.10.	Verify windshield wiper and washer operation; replace wiper blades.	Priority Rating 1

Master Automobile Service Technology (MAST) Task List

A. General Electrical System Diagnosis

A.1.	Research applicable vehicle and service information, vehicle service history, service precautions, and technical service bulletins.	Priority Rating 1
A.2.	Demonstrate knowledge of electrical/electronic series, parallel, and series-parallel circuits using principles of electricity (Ohm's law).	Priority Rating 1
A.3.	Demonstrate proper use of a digital multimeter (DMM) when measuring source voltage, voltage drop (including grounds), current flow, and resistance.	Priority Rating 1
A.4.	Demonstrate knowledge of the causes and effects from shorts, grounds, opens, and resistance problems in electrical/electronic circuits.	Priority Rating 1
A.5.	Check operation of electrical circuits with a test light.	Priority Rating 1
A.6.	Check operation of electrical circuits with fused jumper wires.	Priority Rating 1

A.7.   Use wiring diagrams during the diagnosis (troubleshooting) of electrical/electronic circuit problems.   Priority Rating 1

A.8.   Diagnose the cause(s) of excessive key-off battery drain (parasitic draw); determine necessary action.   Priority Rating 1

A.9.   Inspect and test fusible links, circuit breakers, and fuses; determine necessary action.   Priority Rating 1

A.10. Inspect and test switches, connectors, relays, solenoid solid state devices, and wires of electrical/electronic circuits; determine necessary action.   Priority Rating 1

A.11. Replace electrical connectors and terminal ends.   Priority Rating 1

A.12. Repair wiring harness.   Priority Rating 3

A.13. Perform solder repair of electrical wiring.   Priority Rating 1

A.14. Check electrical/electronic circuit waveforms; interpret readings and determine needed action.   Priority Rating 2

A.15. Repair CAN/BUS wiring harness.   Priority Rating 1

B.   Battery Diagnosis and Service

B.1.   Perform battery state-of-charge test (conductance); determine necessary action.   Priority Rating 1

B.2.   Confirm proper battery capacity for vehicle application; perform battery capacity test; determine necessary action.   Priority Rating 1

B.3.   Maintain or restore electronic memory functions.   Priority Rating 1

B.4.   Inspect and clean battery; fill battery cells; check battery cables, connectors, clamps, and hold-downs.   Priority Rating 1

B.5.   Perform slow/fast battery charge according to manufacturer's recommendations.   Priority Rating 1

B.6.   Jump-start vehicle using jumper cables and a booster battery or an auxiliary power supply.   Priority Rating 1

B.7.   Identify high-voltage circuits of electric or hybrid electric vehicle and related safety precautions.   Priority Rating 3

B.8.   Identify electronic modules, security systems, radios, and other accessories that require reinitialization or code entry after reconnecting vehicle battery.   Priority Rating 1

B.9.   Identify hybrid vehicle auxiliary (12v) battery service, repair, and test procedures.   Priority Rating 3

C.   Starting System Diagnosis and Repair

C.1.   Perform starter current draw test; determine necessary action.   Priority Rating 1

C.2.   Perform starter circuit voltage drop tests; determine necessary action.   Priority Rating 1

C.3.   Inspect and test starter relays and solenoids; determine necessary action.   Priority Rating 2

C.4.   Remove and install starter in a vehicle.   Priority Rating 1

C.5.   Inspect and test switches, connectors, and wires of starter control circuits; determine necessary action.   Priority Rating 2

C.6.   Differentiate between electrical and engine mechanical problems that cause a slow-crank or a no-crank condition.   Priority Rating 2

D.   Charging System Diagnosis and Repair

D.1.   Perform charging system output test; determine necessary action.   Priority Rating 1

D.2.   Diagnose (troubleshoot) charging system for causes of undercharge, no-charge, or overcharge conditions.   Priority Rating 1

D.3.   Inspect, adjust, or replace generator (alternator) drive belts; check pulleys and tensioners for wear; check pulley and belt alignment.   Priority Rating 1

D.4.   Remove, inspect, and reinstall generator (alternator).   Priority Rating 1

D.5.   Perform charging circuit voltage drop tests; determine necessary action.   Priority Rating 1

E. Lighting Systems Diagnosis and Repair
E.1.   Diagnose (troubleshoot) the causes of brighter-than-normal, intermittent, dim, or no light operation; determine necessary action.   Priority Rating 1
E.2.   Inspect interior and exterior lamps and sockets including headlights and auxiliary lights (fog lights/driving lights); replace as needed.   Priority Rating 1
E.3.   Aim headlights.   Priority Rating 2
E.4.   Identify system voltage and safety precautions associated with high intensity discharge headlights.   Priority Rating 2

F. Gauges, Warning Devices, and Driver Information Systems Diagnosis and Repair
F.1.   Inspect and test gauges and gauge sending units for causes of abnormal gauge readings; determine necessary action.   Priority Rating 2
F.2.   Diagnose (troubleshoot) the causes of incorrect operation of warning devices and other driver information systems; determine necessary action.   Priority Rating 2

G. Horn and Wiper/Washer Diagnosis and Repair
G.1.   Diagnose (troubleshoot) causes of incorrect horn operation; perform necessary action.   Priority Rating 1
G.2.   Diagnose (troubleshoot) causes of incorrect wiper operation; diagnose wiper speed control and park problems; perform necessary action.   Priority Rating 2
G.3.   Diagnose (troubleshoot) windshield washer problems; perform necessary action.   Priority Rating 2

H. Accessories Diagnosis and Repair
H.1.   Diagnose (troubleshoot) incorrect operation of motor-driven accessory circuits; determine necessary action.   Priority Rating 2
H.2.   Diagnose (troubleshoot) incorrect electric lock operation (including remote keyless entry); determine necessary action.   Priority Rating 2
H.3.   Diagnose (troubleshoot) incorrect operation of cruise control systems; determine necessary action.   Priority Rating 3
H.4.   Diagnose (troubleshoot) supplemental restraint system (SRS) problems; determine necessary action.   Priority Rating 2
H.5.   Disable and enable airbag system for vehicle service; verify indicator lamp operation.   Priority Rating 1
H.6.   Remove and reinstall door panel.   Priority Rating 1
H.7.   Check for module communication errors (including CAN/BUS systems) using a scan tool.   Priority Rating 2
H.8.   Describe the operation of keyless entry/remote-start systems.   Priority Rating 3
H.9.   Verify operation of instrument panel gauges and warning/indicator lights; reset maintenance indicators.   Priority Rating 1
H.10. Verify windshield wiper and washer operation; replace wiper blades.   Priority Rating 1
H.11. Diagnose (troubleshoot) radio static and weak, intermittent, or no radio reception; determine necessary action.   Priority Rating 3
H.12. Diagnose (troubleshoot) body electronic system circuits using a scan tool; determine necessary action.   Priority Rating 3
H.13. Diagnose the cause(s) of false, intermittent, or no operation of anti-theft systems.   Priority Rating 3
H.14. Perform software transfers, software updates, or flash reprogramming on electronic modules.   Priority Rating 3

DEFINITIONS OF TERMS USED IN THE TASK LIST

To clarify the intent of these tasks, NATEF has defined some of the terms used in the task listings. To get a good understanding of what the task includes, refer to this glossary while reading the task list.

adjust	To bring components to specified operational settings.
align	To restore the proper position of components.
analyze	Assess the condition of a component or system.
assemble (reassemble)	To fit together the components of a device or system.
charge	To bring to a specified state (e.g., battery or air conditioning system).
check	To verify condition by performing an operational or comparative examination.
clean	To rid components of foreign matter for the purpose of reconditioning, repairing, measuring, and reassembling.
confirm	To acknowledge something has happened with firm assurance.
Controller Area Network (CAN)	CAN is a network protocol (SAE J2284/ISO 15765-4) used to interconnect a network of electronic control modules.
demonstrate	To show or exhibit the knowledge of a theory or procedure.
determine	To establish the procedure to be used to perform the necessary repair.
determine necessary action	Indicates that the diagnostic routine(s) is the primary emphasis of a task. The student is required to perform the diagnostic steps and communicate the diagnostic outcomes and corrective actions required, addressing the concern or problem. The training program determines the communication method (worksheet, test, verbal communication, or other means deemed appropriate) and whether the corrective procedures for these tasks are actually performed.
diagnose	To identify the cause of a problem.
differentiate	To perceive the difference in or between one thing to other things.
disassemble	To separate a component's parts as a preparation for cleaning, inspection, or service.
discharge	To empty a storage device or system.
high voltage	Voltages of 50 volts or higher.
identify	To establish the identity of a vehicle or component prior to service; to determine the nature or degree of a problem.
inspect	(see *check*)
install (reinstall)	To place a component in its proper position in a system.
jump-start	To use an auxiliary power supply to assist a battery to crank an engine.
listen	To use audible clues in the diagnostic process; to hear the customer's description of a problem.
locate	Determine or establish a specific spot or area.
maintain	To keep something at a specified level, position, rate, etc.
measure	To compare existing dimensions/values for comparison to specifications.
mount	To attach or place a tool or component in the proper position.
network	A system of interconnected electrical modules or devices.
on-board diagnostics (OBD)	Diagnostic protocol that monitors computer inputs and outputs for failures.

parasitic draw   Electrical loads that are still present when the ignition circuit is turned OFF.

perform   To accomplish a procedure in accordance with established methods and standards.

perform necessary action Indicates that the student is to perform the diagnostic routine(s) and perform the corrective action item. Where various scenarios (conditions or situations) are presented in a single task, at least one of the scenarios must be accomplished.

priority ratings   Indicates the minimum percentage of tasks, by area, a program must include in its curriculum in order to be certified in that area.

reassemble   (see *assemble*)

remove   To disconnect and separate a component from a system.

repair   To restore a malfunctioning component or system to operating condition.

replace   To exchange a component; to reinstall a component.

research   To make a thorough investigation into a situation or matter.

reset   (see *set*)

service   To perform a specified procedure as specified in the owner's or service manual.

set   To adjust a variable component to a given, usually initial, specification.

test   To verify a condition through the use of meters, gauges, or instruments.

verify   To confirm that a problem exists after hearing the customer's concern; or, to confirm the effectiveness of a repair.

voltage drop   A reduction in voltage (electrical pressure) caused by the resistance in a component or circuit.

CROSS-REFERENCE GUIDES

RST Task	Job Sheet
Shop and Personal Safety	
1.	1
2.	2
3.	2
4.	2
5.	2
6.	2
7.	3
8.	1
9.	2
10.	1
11.	1
12.	1
13.	4, 5
14.	6
15.	7
Tools and Equipment	
1.	8
2.	8
3.	8
4.	8
5.	8
Preparing Vehicle for Service	
1.	9
2.	9
3.	9
4.	9
5.	9
Preparing Vehicle for Customer	
1.	9

MLR Task	Job Sheet
A.1	10
A.2	11
A.3	12
A.4	13

MLR Task	Job Sheet
A.5	14
A.6	13
A.7	15
A.8	16
A.9	17
A.10	20
A.11	19
B.1	23
B.2	23
B.3	24
B.4	25
B.5	26
B.6	27
B.7	28
B.8	29
B.9	30
C.1	31
C.2	31
C.3	31
C.4	32
C.5	33
D.1	35
D.2	36
D.3	37
D.4	35
E.1	38
E.2	39
E.3	40
F.1	50
F.2	51
F.3	53
F.4	54
F.5	45, 55

AST Task	Job Sheet
A.1	10
A.2	11
A.3	12
A.4	13
A.5	14
A.6	13
A.7	15
A.8	16
A.9	17
A.10	18
A.11	19
A.12	19
A.13	20
B.1	23
B.2	23
B.3	24
B.4	25

AST Task	Job Sheet
B.5	26
B.6	27
B.7	28
B.8	29
B.9	30
C.1	31
C.2	31
C.3	31
C.4	32
C.5	33
C.6	34
D.1	35
D.2	35
D.3	36
D.4	37
D.5	35
E.1	38
E.2	38
E.3	39
E.4	40
F.1	41
F.2	42
G.1	43
G.2	44
G.3	45
H.1	46
H.2	47
H.3	48
H.4	49
H.5	50
H.6	51
H.7	52
H.8	53
H.9	54
H.10	55

MAST Task	Job Sheet
A.1	10
A.2	11
A.3	12
A.4	13
A.5	14
A.6	13
A.7	15
A.8	16
A.9	17
A.10	18
A.11	19
A.12	19
A.13	20
A.14	21
A.15	22

MAST Task	Job Sheet
B.1	23
B.2	23
B.3	24
B.4	25
B.5	26
B.6	27
B.7	28
B.8	29
B.9	30
C.1	31
C.2	31
C.3	31
C.4	32
C.5	33
C.6	34
D.1	35
D.2	35
D.3	36
D.4	37
D.5	35
E.1	38
E.2	38
E.3	39
E.4	40
F.1	41
F.2	42
G.1	43
G.2	44
G.3	45
H.1	46
H.2	47
H.3	48
H.4	49
H.5	50
H.6	51
H.7	52
H.8	53
H.9	54
H.10	55
H.11	56
H.12	57
H.13	58
H.14	59

PREFACE

The automotive service industry continues to change with the technological changes made by automobile, tool, and equipment manufacturers. Today's automotive technician must have a thorough knowledge of automotive systems and components, good computer skills, exceptional communication skills, good reasoning, the ability to read and follow instructions, and above average mechanical aptitude and manual dexterity.

This new edition, like the last, was designed to give students a chance to develop the same skills and gain the same knowledge that today's successful technician has. This edition also reflects the changes in the guidelines established by the National Automotive Technicians Education Foundation (NATEF) in 2013.

The purpose of NATEF is to evaluate technician training programs against standards developed by the automotive industry and recommend qualifying programs for certification (accreditation) by ASE (National Institute for Automotive Service Excellence). Programs can earn ASE certification upon the recommendation of NATEF. NATEF's national standards reflect the skills that students must master. ASE certification through NATEF evaluation ensures that certified training programs meet or exceed industry-recognized, uniform standards of excellence.

At the expense of much time and thought, NATEF has assembled a list of basic tasks for each of their certification areas. These tasks identify the basic skills and knowledge levels that competent technicians have. The tasks also identify what is required for a student to start a successful career as a technician.

In June 2013, after many discussions with the industry, NATEF established a new model for automobile program standards. This new model is reflected in this edition and covers the new standards, which are based on three (3) levels: Maintenance & Light Repair (MLR), Automobile Service Technician (AST), and Master Automobile Service Technician (MAST). Each successive level includes all the tasks of the previous level, in addition to new tasks. In other words, the AST task list includes all of the MLR tasks, plus additional tasks. The MAST task list includes all of the AST tasks, plus additional tasks specifically used for MAST.

Most of the content in this book are job sheets. These job sheets relate to the tasks specified by NATEF, according to the appropriate certification level. The main considerations during the creation of these job sheets were student learning and program certification by NATEF. Students are guided through standard industry-accepted procedures. While they are progressing, they are asked to report their findings, as well as offer their thoughts on the steps they have just completed. The questions asked of students are thought provoking and require students to apply what they know to what they observe.

The job sheets were also designed to be generic. That is, whenever possible, the tasks can be performed on any vehicle from any manufacturer. Also, completion of the sheets does not require the use of specific brands of tools and equipment; rather, students use what is available. In addition, the job sheets can be used as a supplement to any good textbook.

Also included are descriptions of tools and equipment listed in NATEF's standards, and explanations of their basic use. The standards recognize that not all programs have the same needs, nor do all programs teach all of the NATEF tasks. Therefore, the basic philosophy for the tools and equipment requirement is that the training should be as thorough as possible with the tools and equipment necessary for those tasks.

Theory instruction and hands-on experience of the basic tasks provide initial training for employment in automotive service or further training in any or all of the specialty areas. Competency in the tasks indicates to employers that you are skilled in that area. You need to know the appropriate theory, safety, and support information for each required task. This should include identification and use of the required tools and testing and measurement equipment required for the tasks, the use of current reference and training materials, the proper way to write work orders and warranty reports, and the storage, handling, and use of hazardous materials as required by the "Right to Know" law, and federal, state, and local governments.

Words to the Instructor: We suggest you grade these job sheets based on completion and reasoning. Make sure the students answer all questions. Then, look at their reasoning to see if the task was actually completed and to get a feel for their understanding of the topic. It will be easy for students to copy others' measurements and findings, but each student should have their own base of understanding and that will be reflected in their explanations.

Words to the Student: While completing the job sheets, you have a chance to develop the skills you need to be successful. When asked for your thoughts or opinions, think about what you observed. Think about what could have caused those results or conditions. You are not being asked to give accurate explanations for everything you do or observe. You are only asked to think. Thinking leads to understanding. Good technicians are good because they have a basic understanding of what they are doing and why they are doing it.

Jack Erjavec and Ken Pickerill

ELECTRICAL AND ELECTRONIC SYSTEMS

To prepare you for what you should learn from completing the job sheets, some basics must be covered. This discussion begins with an overview of electrical and electronic systems. Emphasis is placed on what they do and how they work. This includes the major components and designs of electrical and electronic systems and their role in the efficient operation of electrical and electronic systems of all designs.

Preparing to work on an automobile would not be complete if certain safety issues were not addressed. This discussion covers those things you should and should not do while working on electrical and electronic systems. Included are proper ways to deal with hazardous and toxic materials.

NATEF's task list for Electrical and Electronic Systems certification is also given with definitions of some of the terms used to describe the tasks. This list gives you a good look at what the experts say you need to know before you can be considered competent to work on electrical and electronic systems.

Following the task list are descriptions of the various tools and types of equipment you need to be familiar with. These are the tools you will use to complete the job sheets. They are also the tools NATEF has identified as being necessary for servicing electrical and electronic systems.

Following the tool discussion is a cross-reference guide that shows which NATEF tasks are related to specific job sheets. In most cases there are single job sheets for each task. Some tasks are part of a procedure and when that occurs, one job sheet may cover two or more tasks. The remainder of the book contains the job sheets.

BASIC ELECTRICAL AND ELECTRONIC SYSTEM THEORY

The electrical and electronic systems are critical parts of all automotive systems. In addition to playing a vital role in starting the vehicle and providing power for lighting and other auxiliary safety systems, these systems must also provide power to the sophisticated controls found on engine, brake, suspension, emission, steering/stability control, and other automotive systems.

A basic understanding of electrical principles is important to properly diagnose any system that is monitored, controlled, or operated by electricity.

Flow of Electricity

All things are made up of atoms and the basics of electricity focus in on atoms. The following principles describe atoms, which are the building blocks of all materials.

- In the center of every atom is a nucleus.
- The nucleus contains positively charged particles called protons and particles called neutrons that have no charge.
- Negatively charged particles called electrons orbit around every nucleus.
- Every type of atom has a different number of protons and electrons, but each atom has an equal number of protons and electrons. Therefore, the total electrical charge of an atom is zero, or neutral.

The looseness or tightness of the electrons in orbit around the neutron of an atom explains the behavior of electricity. Electricity is caused by the flow of electrons from one atom to another. The release of energy as one electron leaves the orbit of one atom and jumps into the orbit of another is electricity. The key behind creating electricity is to give a reason for the electrons to move.

There is a natural attraction of electrons to protons. Electrons have a negative charge and are attracted to something with a positive charge. When an electron leaves the orbit of an atom, the atom then has a positive charge. An electron moves from one atom to another because the atom next to it appears to be more positive than the one it is orbiting around. An electrical power source provides for a more positive charge, and to allow for a continuous flow of electricity, it supplies free electrons. To have a continuous flow of electricity, three things must be present: an excess of electrons in one place, a lack of electrons in another place, and a path between the two places.

Two power or energy sources are used in an automobile's electrical system. These are based on a chemical reaction and on magnetism. A car's battery (Figure 1) is a source of chemical energy. A chemical reaction in the battery provides for an excess of electrons and a lack of electrons in another place. Batteries have two terminals: a positive and a negative. Basically, the negative terminal is the outlet for the electrons and the positive terminal is the inlet for the electrons to get to the protons.

The chemical reaction in a battery causes a lack of electrons at the positive (+) terminal and excess at the negative (−) terminal. This creates an electrical imbalance, causing the electrons to flow through the path provided by a wire.

The chemical process in the battery continues to provide electrons until the chemicals become weak. At that time, either the battery has run out of free electrons or each of the protons is matched with an electron. When this happens, there is no longer a reason for the electrons to move to the positive side of the battery. Fortunately, the vehicle's charging system frees up electrons in the battery and allows the chemical reaction in the battery to continue indefinitely.

Electricity and magnetism are interrelated. One can be used to produce the other. Moving a wire through an already existing magnetic field can produce electricity. This process of producing electricity through magnetism is called induction. In an AC generator, a magnetic field is moved through a coil of wire and voltage is induced. The amount of electricity produced depends on a number of factors, including the strength of the magnetic field, the number of wires that pass through the field, and the speed at which the wire moves through the magnetic field.

Electrical Terms

Electrical current is a term used to describe the movement or flow of electricity. The greater the number of electrons flowing past a given point in a given amount of time, the more current the circuit has. This current, like the flow of water or any other substance, can be measured. Voltage is electrical pressure. Voltage is the force developed by the attraction of the electrons to the protons. The more positive one side of the circuit is, the more voltage is present in the circuit. Voltage does not flow; rather, it is the pressure that causes current flow.

When any substance flows, it meets resistance. The resistance to electrical flow can be measured. Power is the rate at which work can be done. Forcing a current through a resistance is work. The power used in a circuit can also be determined.

Electrical Current

The unit for measuring electrical current is the ampere. There are two types of electrical flow, or current: direct current (DC) and alternating current (AC). In direct current, the electrons flow in one direction only. In alternating current, the electrons change direction at a fixed rate. An automobile uses DC current, whereas the current in homes and buildings is AC.

Figure 1 A typical battery.

Voltage

In electrical flow, some force is needed to move the electrons between atoms. This force is the pressure that exists between a positive and negative point within an electrical circuit. This force, also called electromotive force (EMF), is measured in units called volts. One volt is the amount of pressure (force) required to move one ampere of current through a resistance of one ohm.

Resistance

In every atom, the electrons resist being moved out of their shell. The amount of resistance depends on the type of atom. In some atoms there is very little resistance to electron flow because the outer electron is loosely held. Materials made up of these types of atoms are typically referred to as conductors. In other substances, there is more resistance to flow, because the outer electrons are tightly held. These materials are called insulators.

The resistance to current flow produces heat. This heat can be measured to determine the amount of resistance. A unit of measured resistance is called an ohm.

Loads

Something in an electrical circuit that has resistance is called a "load." Loads can be a desirable resistance or a nondesirable one. In either case, a load consumes electrical energy. The energy used by a load is measured in volts. Amperage stays constant in a circuit, but the voltage is dropped as it powers a load. Measuring voltage drop determines the amount of energy consumed by the load.

Ohm's Law

Much of the understanding of the behavior of electricity is based on Ohm's Law. Ohm's Law is a statement and definition about the relationship between voltage, current, and resistance. This relationship is expressed in a mathematical formula that says there must be one volt of electrical pressure to push one ampere of electrical current through one ohm of electrical resistance. With this formula, the behavior of electricity can be predicted.

We know, from Ohm's Law, that when resistance in a circuit increases, circuit current decreases. And when circuit resistance decreases, current increases. Therefore, the resistance in that circuit determines the amount of current that flows in a circuit.

Circuits

When electrons are able to flow along a path (wire) between two points, an electrical circuit is formed. An electrical circuit is considered complete when there is a path that connects the positive and negative terminals of the electrical power source. Somewhere in the circuit there must be a load or resistance to control the amount of current in the circuit. Most automotive electrical circuits use the chassis as the path to the negative side of the battery. Electrical components have a lead that connects them to the chassis. These are called the chassis ground connections.

The metal frame acts as the return wire in the circuit. Current passes from the battery, through the load, and into the frame. The frame is connected to the negative terminal of the battery through the battery's ground wire. This completes the circuit. An electrical component, such as a generator, is often mounted directly to the engine block, transmission case, or frame.

This direct mounting effectively grounds the component without the use of a separate ground wire. In other cases, however, a separate ground wire must be run from the component to the frame or another metal part to ensure a sound return path. The increased use of plastics and other nonmetallic materials in body panels and engine parts has made electrical grounding more difficult. To assure good grounding back to the battery, some manufacturers now use a network of common grounding terminals and wires.

In a complete circuit, the flow of electricity can be controlled and applied to do useful work, such as to cause a headlamp to light or turn over a starter motor. Components that use electrical power put a load on the circuit and change electrical energy into another form of energy, such as heat energy.

Most automotive circuits contain four basic parts: (1) power sources, such as a battery or generator that provides the energy needed to create electron flow; (2) conductors, such as copper wires that provide a path for current flow; (3) loads, which are devices that use electricity to perform work, such as light bulbs, electric motors, or resistors; and (4) controllers, such as switches or relays that direct the flow of electrons.

Conductors and Insulators

Controlling and routing the flow of electricity requires the use of materials known as conductors and insulators. Conductors are materials with a low resistance to the flow of current. If the number of electrons in the outer shell or ring of an atom is less than 4, the force holding them in place is weak. The voltage needed to move these electrons and create current flow is relatively small. Most metals, such as copper, silver, and aluminum are excellent conductors.

When the number of electrons in the outer ring is greater than 4, the force holding them in orbit is very strong, and very high voltages are needed to dislodge them. These materials are known as insulators. They resist the flow of current. Thermal plastics are the most common electrical insulators used today. They can resist heat, moisture, and corrosion without breaking down.

Resistors

Automotive electrical circuits contain a number of different types of electrical devices. Resistors are used to limit current flow (and thereby voltage) in circuits where full current flow and voltage are not needed. Resistors are devices specially constructed to introduce a measured amount of electrical resistance into a circuit. In addition, some other components use resistance to produce heat and even light. An electric window defroster is a specialized type of resistor that produces heat. Electric lights are resistors that get so hot they produce light.

Three different types of resistors are found in automobiles: fixed value, stepped or tapped, and variable. Fixed value resistors are designed to have only one rating, which should not change. These resistors are used to control voltage such as in an automotive ignition system. Tapped or stepped resistors are designed to have two or more fixed values, available by connecting wires to the several taps of the resistor. Blower motor resistor packs, which provide for different fan speeds, are an example of this type of resistor. Variable resistors are designed to have a range of resistances available through two or more taps and a control.

Three commonly used variable resistors are rheostats, potentiometers, and thermistors. Rheostats have two connections, one to the fixed end of a resistor and one to a sliding contact with the resistor. Turning the control moves the sliding contact away from or toward the fixed end tap, increasing or decreasing the resistance. Potentiometers have three connections, one at each end of the resistance and one connected to a sliding contact with the resistor. Turning the

control moves the sliding contact away from one end of the resistance, but toward the other end.

Thermistors are designed to change their resistance values in response to changes in temperature. Thermistors are used to provide compensating voltage in components or to determine temperature. As a temperature sender, the thermistor is connected to a voltmeter calibrated in degrees. As the temperature rises or falls, the resistance also changes. This changes the reading on the meter.

Circuit Protection Devices

When overloads or shorts in a circuit allow too much current to flow, the wiring in the circuit heats up, the insulation melts, and a fire can result, unless the circuit has some kind of protective device. Fuses, fuse links, maxi-fuses, and circuit breakers are designed to provide protection from high current. They may be used singly or in combination.

Fuses are rated according to the current at which they are designed to open or blow. The current rating for blade fuses is indicated by the color of the plastic case. In addition, it is usually marked on the top. Fuse or fusible links are used in circuits where maximum current controls are not so critical. They are often installed in the positive battery lead that powers the ignition switch and other circuits that are live with the key off. A fuse link is a short length of small-gauge wire installed in a conductor. Because the fuse link is a lighter gauge of wire than the main conductor, it melts and opens the circuit before damage can occur in the rest of the circuit.

Often a maxi-fuse is used instead of a fusible link. Maxi-fuses look and operate like two-prong, blade or spade fuses, except they are much larger and can handle more current.

Circuit Breakers

Some circuits are protected by circuit breakers, which, like fuses, are rated in amperes. A circuit breaker conducts current through an arm made of two types of metal bonded together (bimetal arm). If the arm starts to carry too much current, it heats up. As one metal expands faster than the other, the arm bends; that opens the contacts and the current flow is broken.

Switches

A switch of some type usually controls electrical circuits. Switches turn the circuit on or off or they are used to direct the flow of current in a circuit. Switches can be under the control of the driver or can be self-operating through a condition of the circuit, the vehicle, or the environment.

Relays

A relay is an electric switch that allows a small amount of current to control a much larger one. When the control circuit switch is open, no current flows to the coil of the relay, so the windings are deenergized. When the switch is closed, the coil is energized, turning the soft iron core into an electromagnet and drawing the armature down. This closes the power circuit contacts, connecting power to the load circuit. When the control switch is opened, the current stops slowing in the coil, the electromagnetic field disappears, and the armature is released, which breaks the power circuit contacts.

Solenoids

Solenoids are also electromagnets with movable cores used to translate electrical current flow into mechanical movement. They can also close contacts, acting as a relay at the same time.

Capacitors (Condensers)

Capacitors are constructed from two or more sheets of electrically conducting material with a nonconducting or dielectric material placed between them and conductors connected to the two sheets. Capacitors are typically used as filters to remove spikes and noise from voltage signals.

Electromagnetism Basics

Electricity and magnetism are related. Current flowing through a wire creates a magnetic field around the wire. Moving a wire through a magnetic field creates current flow in the wire. Many automotive components, such as generators, ignition coils, starter solenoids, and magnetic pulse generators operate using principles of electromagnetism.

A magnet has two points of maximum attraction, one at each end of the magnet. These points are called poles, with one being designated the North pole and the other the South pole. When two magnets are brought together, opposite poles attract, while similar poles repel each other.

If a straight piece of conducting wire is moved across a magnetic field and the terminals of a voltmeter is attached to both ends of the wire, the voltmeter will show a small voltage reading. A voltage has been induced in the wire. The conducting wire is cutting across the flux lines of the magnetic field and is inducing a voltage.

As a result of this induction process, the wire or conductor becomes a source of electricity and has a polarity or distinct positive and negative end. However, this polarity can be switched depending on the relative direction of movement between the wire and magnetic field. That is why an AC generator produces alternating current. Moving the wire parallel to the lines of flux does not induce voltage.

Basics of Electronics

Simply put, electronics is a technology used to control electricity. Electronics has become a special technology beyond electricity. Transistors, diodes, semiconductors, integrated circuits, and solid-state devices are all considered to be part of electronics rather than just electrical devices. But keep in mind that all the basic laws of electricity apply to electronic controls.

A semiconductor is a material or device that can function as either a conductor or an insulator, depending on how its structure is arranged. Semiconductor materials have less resistance than an insulator but more resistance than a conductor. Because semiconductors have no moving parts, they seldom wear out or need adjustment. Semiconductors are also small, require little power to operate, are reliable, and generate very little heat. For all these reasons, semiconductors are being used in many applications.

Because a semiconductor can function as both a conductor and an insulator, it is very useful as a switching device. How a semiconductor functions depends upon the way current flows (or tries to flow) through it. Two common semiconductor devices are diodes and transistors.

Diodes

The diode is the simplest semiconductor device. A diode allows current to flow in one direction, but not in the opposite direction. Therefore, it can function as a switch, acting as either a conductor or insulator, depending on the direction of current flow. The most commonly used diodes are regular diodes, LEDs, zener diodes, clamping diodes, and photo diodes.

One application of diodes is in the AC generator, where they function as one-way valves for current flow. Generators produce alternating current. The AC must be changed to DC before it is sent to the vehicle's electrical circuit. Diodes in the generator are arranged so that current can leave the generator in one direction only (as direct current).

A variation of the diode is the zener diode. This device functions like a standard diode until a certain voltage is reached. When the voltage level reaches this point, the zener diode allows current to flow in the reverse direction. Zener diodes are often used in electronic voltage regulators.

Whenever the current flow through a coil of wire (such as used in a solenoid or relay) stops, a voltage surge or spike is produced. This surge results from the collapsing of the magnetic field around the coil. The movement of the field across the winding induces a very high voltage spike, which can damage electronic components. In the past, a capacitor was used as a "shock absorber" to prevent component damage from this surge. On today's vehicles, a clamping diode is commonly used to prevent this voltage spike. By installing a clamping diode in parallel to the coil, a bypass is provided for the electrons during the time the circuit is opened.

An example of the use of clamping diodes is on some air-conditioning compressor clutches. Because the clutch operates by electromagnetism, opening the clutch coil circuit produces a voltage spike. If the spike was left unchecked, it could damage the clutch coil relay contacts or the vehicle's computer.

Transistors

A transistor is an electronic device produced by joining three sections of semiconductor materials. Like the diode, it is very useful as a switching device, functioning as either a conductor or an insulator. Transistors are also used as signal amplifiers in radios, stereos, calculators, computers, and computerized engine controls.

Integrated Circuits

An integrated circuit is simply a large number of diodes, transistors, and other electronic components such as resistors and capacitors, all mounted on a single piece of semiconductor material. These circuits are very small and technology is

allowing them to be made smaller and more complex. The increasingly small size of integrated circuits is very important to automobiles.

Integrated circuits are the building blocks of the modules in a vehicle's networks which have provided us with more safety, reliability, fuel economy and lower emissions.

Electronic Circuits

A typical electronic control system is made up of sensors, actuators, and related wiring that is tied into a central processor called a microprocessor or computer. Most input sensors are designed to produce a voltage signal that varies within a given range (from high to low, including all points in between). A signal of this type is called an analog signal. Unfortunately, the computer doesn't understand analog signals. It can only read a digital binary signal, which is a signal that has only two values—on or off.

To overcome this communication problem, all analog voltage signals are converted to a digital format by an analog-to-digital converter (A/D converter). Some sensors like the Hall-effect switch produce a digital or square wave signal that can go directly to the microcomputer as input. The term **square wave** is used to describe the appearance of a digital circuit after it has been plotted on a graph. The abrupt changes in circuit condition (on and off) result in a series of horizontal and vertical lines that connect to form a square-shaped pattern.

In addition to A/D conversion, some voltage signals require amplification before they can be relayed to the computer. To perform this task, an input conditioner known as an amplifier is used to strengthen weak voltage signals.

After input has been generated, conditioned, and passed along to the microcomputer, it is ready to be processed for the purposes of performing work and displaying information. The portion of the microcomputer that receives sensor input and handles all calculations (makes decisions) is called the microprocessor. In order for the microprocessor to make the most informed decisions regarding system operation, sensor input is supplemented by the memory.

A computer's memory holds the programs and other data, such as vehicle calibrations, that the microprocessor refers to while performing calculations. To the computer, the program is a set of instructions or procedures that it must follow. Included in the program is information that tells the microprocessor when to retrieve input (based on temperature, time, etc.), how to process the input, and what to do with it once it has been processed.

The microprocessor works with memory in two ways: it can read information from memory or change information in memory by writing in or storing new information. During processing, the computer often receives more data than it can immediately handle. In these instances, some information is temporarily stored or written into memory until the microprocessor needs it.

Sensors

All sensors perform the same basic function. They detect a mechanical condition (movement or position), chemical state, or temperature condition and change it into an electrical signal that can be used by the computer to make decisions. The computer makes decisions based on information it receives from sensors and the programmed instructions it has in its memory. Each sensor used in a particular system has a specific job to do (for example, monitor throttle position, vehicle speed, and manifold pressure). Together these sensors provide enough information to help the computer form a complete picture of vehicle operation.

Reference voltage (Vref) sensors provide input to the computer by modifying or controlling a constant, predetermined voltage signal. This signal, which can have a reference value from 5 to 9 volts, is generated and sent out to each sensor by a reference voltage regulator located inside the processor. Because the computer knows that a certain voltage value has been sent out, it can indirectly interpret things like motion, temperature, and component position, based on what comes back. Variable resistors are typically used as voltage-reference sensors.

Two other commonly used reference voltage sensors are switches and thermistors. Switches tell the computer when something is turned on or off and when a particular condition exists. Switches don't provide a variable signal; there is either a signal from them, or there is no signal.

Thermistors are temperature sensitive and send a varying signal to the computer based on the temperature they are subject to. Voltage generating sensors include components like the Hall-effect switch, oxygen sensor (zirconium dioxide), and knock sensor (piezoelectric), which are capable of producing their own input voltage signal. This varying voltage signal, when received by the computer, enables the computer to monitor and adjust for changes in the operation of various systems of the automobile.

Air/Fuel Ratio Sensor

Similar to an oxygen sensor, an air fuel ratio sensor has been developed to provide a more precise measurement of the fuel mixture. The A/F ratio sensor produced a current, and the current produced can go either direction, depending on whether the mixture is rich or lean.

Actuators

After the computer has assimilated the information and the tools used by it to process this information, it sends output signals to control devices called actuators. These actuators are solenoids, switchers, relays, or motors, which physically act or carry out a decision the computer has made.

Actuators are electromechanical devices that convert an electrical current into mechanical action. This mechanical action can then be used to open and close valves, control vacuum to other components, or open and close switches. When the computer receives an input signal indicating a change in one or more of the operating conditions, the computer determines the best strategy for handling the conditions. The computer then controls a set of actuators to achieve a desired effect or strategy goal. In order for the computer to control an actuator, it must rely on a component called an output driver.

Output drivers are located in the processor and operate by the digital commands issued by the computer. Basically, the output driver is nothing more than an electronic on/off switch that the computer uses to control the ground or power circuit of a specific actuator.

For actuators that cannot be controlled by a solenoid, relay, switches, or motors; the computer must turn its digitally coded instructions back into an analog format via a digital-to-analog converter.

Multiplexing

Multiplexing is an in-vehicle networking system used to transfer data between electronic modules through a serial data bus. Serial data is electronically coded information that is transmitted by one computer and received and displayed by another computer. Serial data is information that is digitally coded and transmitted in a series of data bits. The data transmission rate is referred to as the baud rate. Baud rate refers to the number of data bits that can be transmitted in a second.

With multiplexing, fewer dedicated wires are required for each function, and this reduces the size of the wiring harness. Using a serial data bus reduces the number of wires by combining the signals on a single wire through time division multiplexing. Information is sent to individual control modules that control each function, such as anti-lock braking, turn signals, power windows, and instrument panel.

Multiplexing also eliminates the need for redundant sensors because the data from one sensor is available to many electronic modules. Multiplexing also allows for greater vehicle content flexibility because functions can be added through software changes, rather than adding another module or modifying an existing one.

The common multiplex system is called the CAN or Controller Area Network. It is used to interconnect a network of electronic control modules.

Batteries

The storage battery is the heart of a vehicle's electrical and electronic systems. It plays an important role in the operation of the starting, charging, ignition, and accessory circuits. The storage battery converts electrical current from the generator into chemical energy, and then stores that energy until it is needed. When switched into an external electrical circuit, the battery's chemical energy is converted back to electrical energy.

A vehicle's battery has three main functions. It provides voltage and serves as a source of current for starting, lighting, and ignition. It acts as a voltage stabilizer for the entire electrical system of the vehicle. And, finally, it provides current whenever the vehicle's electrical demands exceed the output of the charging system.

Battery Voltage and Capacity

The open-circuit voltage of a fully charged battery cell is roughly 2.1 volts in each of its 6 cells. Therefore a fully charged battery has at least 12.6-volts. Cell size, state of charge, rate of discharge, battery condition and design, and electrolyte temperature all strongly influence the voltage of a battery during discharge. When

cranking an engine over at 80°F, the voltage of an average battery may be about 11.5 to 12 volts. At 0°F, the voltage is significantly lower.

The concentration of acid in the electrolyte in the pores of the plates also affects battery voltage or discharge. As the acid chemically combines with the active materials in the plates and is used up, the voltage drops unless fresh acid from outside the plate moves in to take its place. As discharging continues, this outside acid becomes weaker, and sulfate saturates the plate material. It then becomes increasingly difficult for the chemical reaction to continue and, as a result, voltage drops to a level no longer effective in delivering sufficient current to the electrical system.

Battery capacity is the ability to deliver a given amount of current over a period of time. It depends on the number and size of the plates used in the cells and the amount of acid used in the electrolyte. Batteries are rated according to reserve capacity and cold-cranking power.

Starting System

The starting system is designed to turn or crank the engine over until it can operate under its own power. To do this, the starter motor receives electrical power from the battery. The starter motor then converts this energy into mechanical energy, which it transmits through the drive mechanism to the engine's flywheel.

A typical starting system has six basic components and two distinct electrical circuits. The components are the battery, ignition switch, battery cables, magnetic switch (either electrical relay or solenoid), starter motor, and the starter safety switch.

The starting system operates with two connected circuits: the starter circuit and the control circuit. The starter circuit carries heavy current flow from the battery to the starter motor by way of the relay or solenoid. The control circuit controls the action of the solenoid and/or relay. The control circuit uses low current to control the high current starter circuit.

The starting circuit requires two or more heavy-gauge cables that attach directly to the battery. One of these cables makes a connection between the battery's negative terminal and a good chassis ground. The other cable connects the battery's positive terminal with the relay or solenoid. On vehicles where the magnetic switch is not mounted directly on the starter motor, the positive cable is actually two cables. One runs from the positive battery terminal to the switch and the second from the switch to the starter motor terminal.

Every starting system contains some type of magnetic switch that enables the control circuit to open and close the starter circuit. A solenoid-actuated starter is the most common starter system used. In this system, the solenoid mounts directly on top of the starter motor. The solenoid uses the electromagnetic field generated by its coil to perform two distinct jobs. It pushes the drive pinion of the starter motor into mesh with the engine flywheel. This is its mechanical function. Secondly, it acts as an electrical relay switch to energize the motor once the drive pinion is engaged. Once the contact points of the solenoid are closed, full current flows from the battery to the starter motor.

The solenoid assembly has two separate windings: a pull-in winding and a hold-in winding. The two windings have approximately the same number of turns but are wound from different size wire. Together these windings produce the electromagnetic force needed to pull the plunger into the solenoid coil. The heavier pull-in windings draw the plunger into the solenoid, while the lighter gauge windings produce enough magnetic force to hold the plunger in this position.

Both windings are energized when the ignition switch is turned to the start position. When the plunger disc makes contact with the solenoid terminals, the pull-in winding is deactivated. At the same time, the plunger contact disc completes the connection between the battery and the starting motor, directing full battery current to the field coils and starter motor armature for cranking power.

As the solenoid plunger moves, the shift fork also pivots on the pivot pin and pushes the starter drive pinion into mesh with the flywheel ring gear. When the starter motor receives current, its armature starts to turn. This motion is transferred through an overriding clutch and pinion gear to the engine flywheel and the engine is cranked.

Starter relays are connected in series with the battery cables to deliver the high current necessary through the shortest possible battery cables. A relay serves as the control for the starter circuit. Often they are used with positive engagement starters that use the magnetic field inside the starter motor to move the drive pinion into the flywheel.

Some vehicles use both a starter relay and a starter motor mounted solenoid. The relay controls current flow to the solenoid, which in turn controls current flow to the starter motor. This reduces the amount of current flowing through the ignition switch. In other words, it takes less current to activate the relay than to activate the solenoid.

Starter Motors

A starting motor is a special type of electric motor designed to operate under great electrical overloads and to produce very high horsepower. All starting motors are generally the same in design and operation. Basically the starter motor consists of a housing, field coils, an armature, a commutator and brushes, and end frames. The main difference between designs is in the drive mechanism used to engage the flywheel.

The starter housing or frame encloses the internal starter components and protects them from damage, moisture, and foreign materials. The housing supports the field coils and forms a path for the magnetism produced by the current passing through the coils.

The field coils and their pole shoes are securely attached to the inside of the iron housing. The field coils and pole shoes are designed to produce strong stationary electromagnetic fields within the starter body as current is passed through the starter. These magnetic fields are concentrated at the pole shoe.

The field coils connect in series with the armature winding through the starter brushes. This design permits all current passing through the field coil circuit to also pass through the armature windings.

The armature is the only rotating component of the starter. When the starter operates, the current passing through the armature produces a magnetic field in each of its conductors. The reaction between the armature's magnetic field and the magnetic fields produced by the field coils causes the armature to rotate. The armature has two main components: the armature windings and the commutator.

The coils connect to each other and to the commutator so that current from the field coils flows through all of the armature windings at the same time. This action generates a magnetic field around each armature winding, resulting in a repulsion force all around the conductor. This repulsion force causes the armature to turn.

The commutator assembly is made up of heavy copper segments separated from each other and the armature shaft by insulation. The commutator segments connect to the ends of the armature windings. Most starter motors have two to six brushes that ride on the commutator segments and carry the heavy current flow from the stationary field coils to the rotating armature windings via the commutator segments.

The control circuit usually consists of an ignition switch connected through normal gauge wire to the battery and the magnetic switch (solenoid or relay). When the ignition switch is turned to the start position, a small amount of current flows through the coil of the magnetic switch, closing it and allowing full current to flow directly to the starter motor. The ignition switch performs other jobs besides controlling the starting circuit. It normally has at least four separate positions: accessory, off, on (run), and start.

The starting safety switch, often called the neutral safety switch, is a normally open switch that prevents the starting system from operating when the transmission is in gear. The safety switch used with an automatic transmission can be either an electrical switch or a mechanical device.

Contact points on the electrical switch are only closed when the shift selector is in park or neutral. The safety switches used with manual transmissions are usually electrical switches activated by the movement of the clutch pedal and/or position of the shift linkage.

Charging Systems

The charging system converts the mechanical energy of the engine to electrical energy. During cranking, the battery supplies all of the vehicle's electrical energy. However, once the engine is running, the charging system is responsible for producing enough energy to meet the demands of all the loads in the electrical system, while also recharging the battery.

In an AC generator, sometimes called an alternator, a spinning magnetic field rotates inside stationary conductors. As the spinning north and south poles of the magnetic field pass the conducting wires, they induce voltage that first flows in one direction and then in the opposite direction. Because automotive electrical systems operate on direct current, this alternating current must be changed or rectified into direct current by diodes.

The rotor assembly consists of a drive shaft, coil, and two pole pieces. A pulley mounted on the shaft end allows the rotor to be spun by a belt driven from the crankshaft pulley. The rotor produces the rotating magnetic field. The magnetic field is generated by passing a small amount of current through the coil windings. As current flows through the coil, the core is magnetized and the pole pieces assume the magnetic polarity of the end of the core that they touch. Thus, one pole piece has a north polarity and the other has a south polarity. The extensions of the pole pieces form the actual magnetic poles.

Current to create the magnetic field is supplied to the coil from one of two sources, the battery, or when the engine is running, the alternator itself. In either case, the current is passed through the voltage regulator before being applied to the coil. The voltage regulator varies the amount of current supplied. Increasing field current to the coil increases the strength of the magnetic field. This, in turn, increases alternator voltage output. The voltage regulator can be contained inside the alternator, or it can be part of the ECM. The ECM can control the voltage regulator directly; this allows the ECM to raise the idle before a large electrical load comes on line, such as the A/C compressor. Decreasing the field voltage to the coil has the opposite effect. Output voltage decreases.

Slip rings and brushes conduct current to the rotor. Two slip rings are mounted directly on the rotor shaft and are insulated from the shaft and each other. Each end of the field coil connects to one of the slip rings. A carbon brush located on each slip ring carries the current to and from the field coil. Current is transmitted from the field terminal of the voltage regulator through the first brush and slip ring to the field coil. Current passes through the field coil and the second slip ring and brush before returning to ground. The stator is the stationary member of the AC generator and is made up of a number of conductors. The stator assembly in most AC generators has three separate windings, each placed in slightly different positions so their electrical pulses are staggered.

The rotor fits inside the stator. A small air gap between the two allows the magnetic field of the rotor to energize all of the stator windings at the same time.

The output from a generator must be regulated. A voltage regulator controls the amount of current produced by the generator and thus the voltage level in the charging circuit. Without a voltage regulator, the battery would be overcharged and the voltage level in the electrical systems would rise to the point where lights would burn out and fuses and fusible links would blow.

The regulation of the charging circuit is accomplished by varying the amount of field current flowing through the rotor. Output is high when the field current is high and low when the field current is low. The operation of the regulator is comparable to that of a variable resistor in series with the field coil. If the resistance the regulator offers is low, the field current is high and if the resistance is high, the field current is low. The amount of resistance in series with the field coil determines the amount of field current, the strength of the rotor's magnetic field, and thus the amount of generator output. The resistance offered by the regulator varies according to charging system demands or needs.

In order to regulate the charging system, the regulator must have system voltage as an input. This voltage is also called sensing system voltage because the regulator is sensing system voltage. The regulator determines the need for charging current according to the level of the sensing voltage. When the sensing voltage is less than the regulator setting, the regulator increases the field current in order to increase the charging current. As sensing voltage rises, a corresponding decrease in field current and system output occurs. Thus, the regulator responds to changes in system voltage by increasing or decreasing charging current.

On a growing number of late-model vehicles, a separate voltage regulator is no longer used. Instead, the voltage regulation circuitry is located in the vehicle's electronic control module.

This type of system does not control rotor field current by acting like a variable resistor. Instead, the computer switches or pulses field current on and off at a fixed frequency of about 400 cycles per second. By varying on-off times, a correct average field current is produced to provide correct alternator output.

Headlights

In a typical headlight system, power is directly supplied by the battery to the headlight portion of the switch and through a fuse in the fuse panel to the remainder of the switch. Although one of the functions of the headlight switch is to control the headlights, the switch has many other functions.

There are many types of headlight switches. The pullout design has three positions: off, park, and head. Pulling a typical switch knob out to the park or first detent (knob catches at stop) illuminates all exterior lights except the headlights. Instrument panel lights are also illuminated in this position. Pulling the Headlamp switch knob out to the head or second detent illuminates the headlights plus all of those lights illuminated in park position. A dimmer rheostat controls the brightness of dash or instrument illumination lights. When the switch knob is rotated completely counterclockwise, the instrument panel illumination is at maximum brightness level. This position may also turn on the courtesy and dome/map lights.

Push button and rotary headlight switches have three positions: off, park, and head. When the park position is selected, power is applied to all circuits except the headlights. When the headlight position is selected, all of those lights illuminated in the park position are on. With these switches, the interior panel lights are controlled at a separate rheostat switch.

Headlights have both a high and low beam. Switching from high beam to low beam is controlled by a dimmer switch. Most vehicles have column mounted dimmer switches that serve many different purposes. These switches may control the wipers, turn signals, and the high and low beam headlight circuits. Some dimmer switches have the additional feature of being able to energize both the high and low beams even if the headlight switch is off. These circuits are usually referred to as flash to pass.

Some vehicles are equipped with an automatic headlight dimmer system. It automatically switches the headlights from high to low beam in response to light from an approaching vehicle, or light from the taillights of a vehicle being overtaken. Major components of the system are a hi-lo beam photocell, an amplifier, a hi-lo beam relay, a dimmer switch, and a hi-lo beam range control.

Vehicles may also have an automatic headlight system that provides light-sensitive automatic on-off control of the headlights. This system consists of a light-sensitive photocell sensor/amplifier assembly and a headlight control relay. Turning the regular headlight switch on overrides the automatic system.

An automatic delayed-exit system keeps the lights on for a preselected period of time after the ignition switch is turned off.

Most late model vehicles have lighting circuits that are controlled through the BCM or a lighting module. The switches are still present on the dash, but they are simply used as an input to the module and the module actually controls the lighting. The lighting module does allow the lighting to be customized, such as having all the lights come on when the key is initially turned off, and setting the delay for the time the lights remain on after leaving the vehicle.

Headlight Bulbs

Headlight systems typically consist of two or four sealed-beam tungsten or halogen headlight bulbs. On a two-headlight system, each headlight has two filaments, a high beam and a low beam. On the four-headlight system, the two outer headlights are for the low-beam headlamps, and the two inner headlights have only the high-beam filament. Some vehicles use headlights that have a separate small light bulb enclosed in a reflector, with a separate lens in front. This type offers better light control and brightness when compared to sealed-beam types.

A halogen light contains a small quartz-glass bulb. Inside the bulb is a fuel filament surrounded by halogen gas. The small, gas-filled bulb fits within a larger metal reflector and lens element. A glass balloon sealed to the metal reflector allows the halogen bulb to be removed without danger of water or dirt damaging the optics within the light.

HID (High Intensity Discharge) headlights produce light with a longer reach and wider distribution on the road. The HID system uses high-voltage arcing across electrodes inside a light bulb to emit light. The light bulb is a metal halide bulb containing xenon gas, mercury, and metal halide. When high voltage (approximately 20,000 volts) is applied to the electrodes inside the bulb, the xenon gas in the bulb emits light. As the temperature in the bulb increases, the mercury in the bulb evaporates and causes arcing in the bulb. As the temperature continues to increase, the metal halide in the mercury arc separates into metal and iodine atoms. The metal atoms emit a high intensity light.

An electronic control unit, located under each headlight, generates the high voltage that is applied to the electrodes of the bulbs to begin the illumination of the bulb. The control module also controls the amperage and voltage to the bulb to immediately provide an optimal amount of light after the bulbs have been turned ON and to enable the bulbs to continue to illuminate. The control module also has a fail-safe function to protect the vehicle and system from the high-voltage if there is a problem in the headlight system.

Other Lighting

The rear light assembly includes the taillights, turn signal/stop/hazard lights/high-mounted stoplights, rear side marker lights, backup lights, and license plate lights.

Flashers are components of both turn and hazard systems. Older flashers contained a temperature-sensitive bimetallic strip and a heating element. The bimetallic strip is connected to one side of a set of contacts. Voltage from the fuse panel is connected to the other side. When the left turn signal switch is activated, current flows through the flasher unit to the turn signal bulbs. This current causes the heating element to emit heat, which in turn causes the bimetallic strip to bend and open the circuit. The absence of current flow allows the strip to cool and again close the circuit. This intermittent on/off interruption of current flow makes all left turn signal lights flash. Operation of the right turn is the same as the operation of the left turn signals.

Modern turn signals are operated electronically, through a module. The module commands the lamps to flash on and off by provided intermittent power to the bulbs that flash. Additionally, many exterior lamps are actually light emitting diodes (LEDS). LEDS function on very low current and illuminate almost instantly when commanded on by the module, which is an advantage with brake lamps. As an additional benefit, LEDs have a very long life, with no filament to burn out.

The stop lights are usually controlled by a stop light switch mounted on the brake pedal arm. In either case, voltage is present at the stop light switch at all times. Depressing the brake pedal causes the stop light switch contacts to close. Current can then flow to the stop light filament of the rear light assembly. These stay illuminated until the brake pedal is released.

Many modern vehicles also use a lighting module or BCM to control the brake lamps as well as the other exterior lamps. The brake lamp switch is used as an input signal to the lighting module. The lighting module then supplies the current to the brake lamps.

When the vehicle is placed in reverse gear, backup lights are turned on to illuminate the area behind the vehicle. Various types of backup light switches are used depending on the type of transmission used on the vehicle. In general, vehicles with a manual transmission have a separate switch. Those with an automatic transmission use a combination neutral start/backup light switch. The combination neutral start/backup light switch used with automatic transmissions is actually two switches combined in one housing. In park or neutral, current from the ignition switch is applied through the neutral start switch to the starting system. In reverse, current from the fuse panel is applied through the backup light switch to the backup lights.

Many vehicles use the lighting module to control the back up lamps as well. The transmission range switch is used to provide the input top the lighting module to turn on the back up lamps.

Light systems normally use one wire to the light, making use of the car body or frame to provide the ground back to the battery. Since many of the manufacturers have gone to plastic socket and mounting plates (as well as plastic body parts) to reduce weight, many lights must now use two wires to provide the ground connection. Some double-filament lights use two hot wires and a third ground wire. That is, double-filament bulbs have two contacts and two wire connections to them if grounded through the base. If not grounded through the base of the bulb, a two-filament bulb has three contacts and three wires connected to it. Single-filament bulbs may be single- or double-contact types. Single-contact types are often grounded through the bulb base, while most double-contact types have two wires: one live and the other a ground.

Instrument Gauges

Gauges provide the driver with a scaled indication of the condition of a system. All gauges (analog or digital) require an input from either a sender or a sensor. Sender or sensor units change electrical resistance in response to changes or movements made by an external component. Movement may be caused by pressure against a diaphragm, heat, or motion of a float as liquid fills a fuel tank.

Electronic speedometers respond to the input of a vehicle speed sensor. This speed signal is also used by other modules in the vehicle, including speed control, ride control module, the engine control module and others. The input to the ECM is from the vehicle speed sensor.

An oil pressure gauge indicates engine oil pressure based on input from a sensor. The sensor's resistance changes with changes in pressure. Some vehicles have an oil pressure switch that causes a warning lamp to light when pressure is low.

A coolant temperature gauge relies on a variable resistor-type sensor, such as a thermistor. Typically, with low coolant temperatures, sender resistance is high and current flow to the gauge is low. Therefore the gauge reads on the cold side. As coolant temperature increases, sender resistance decreases and current flow increases. The needle moves toward the hot end of the gauge. Most late model vehicles receive the temperature sensor reading from the engine coolant temperature sensor (ECT) that is sent to the ECM or BCM. The signal is then transmitted over the serial data bus to the necessary modules, including the instrument panel control module or IPC. Fuel level gauges rely on a float assembly in the fuel tank to monitor fuel level. The fuel sender unit may be combined with the fuel pump assembly. Typically, the sending unit is a variable resistor controlled by the level of an attached float in the fuel tank. When the fuel level is low, resistance in the sender is low and movement of the gauge indicator dial is minimal (from the empty position). When the fuel level is high, the resistance in the sender is high and movement of the gauge indicator (from the empty position) is greater.

A tachometer indicates engine revolutions per minute (rpm). The tachometer generally receives its signals directly from the ECM.

Windshield Wiper/Washer Systems

There are several types of windshield wiper systems. The wiper systems function to keep the windshield clear of rain, snow, and dirt. Headlight wipers are available that work in unison with conventional windshield wipers. The major components of a wiper system are the control switch, wiper motor and switch, and washer fluid pump. On systems with interval wipers, an interval governor is added to the circuit. The wiper motor produces a rotational motion, and an assembly of levers connected to an offset motor drive changes the rotational motion of the motor to an oscillation motion, which is needed for the wiper blades.

When the wiper switch is moved to one of the on positions, voltage from the fuse panel is applied through the wiper switch directly to the wiper motor on systems without interval wipers. Shutting off the switch cuts power to the wiper motor. However, the wiper motor park switch maintains power to the motor until it is in the park position.

On vehicles with interval wipers, current passes through the interval governor to the wiper motor. When the wiper switch is placed in the interval position, the signal is applied to an interval timer circuit in the governor. The time period of the interval is adjusted by a potentiometer in the wiper switch.

The washer pump is operated by holding the washer switch in the activate position. On models with interval wipers this signal is also applied to the governor. If the wiper switch is in off, an interval override circuit in the governor causes the wipers to operate at low speed until the washer switch is released. Then the wipers operate for several more cycles and park or return to interval operation. The wiper arms and blades are attached directly to the two pivot points operated by the linkage and motor.

Power Door Lock Systems

Although systems for automatically locking doors vary from one vehicle to another, the overall purpose is the same—to lock all outside doors. There are, however, several variations of door arrangements used that require slight differences in components from one system to another. As a safety precaution against being locked in a car due to an electrical failure, power locks can be manually operated.

When either the driver's or passenger's control switch is activated (either locked or unlocked), power from the fuse panel is applied through the switch to the door lock actuator motor. A rod that is part of the actuator moves up or down in the door latch assembly as required to lock or unlock the door. On some models the signal from the switch is applied to a relay that, when energized, applies an activating voltage to the door lock actuator. The door lock actuator consists of a motor and a built-in circuit breaker.

The power door locks on many late model vehicles are controlled by the BCM, through a door module.

Power Trunk Release

The power trunk release system is a relatively simple electrical circuit that consists of a switch and a solenoid. When the trunk release switch is pressed, voltage is applied through the switch to the solenoid. With battery voltage on one side and ground on the other, the trunk release solenoid energizes and the trunk latch releases to open the trunk lid.

Power Windows

Obviously, the primary function of any power window system is to raise and lower windows. The systems do not vary significantly from one model to another. The major components of a typical system are the master control switch, individual window control switches, and the window drive motors. In addition, on four-door models, a window safety relay and in-line circuit breaker are also included. The master control switch provides overall system control.

Four-door model master control switches usually have four segments while two-door models have two segments. Each segment controls power to a separate window motor. Each segment actually operates as a separate independent switch. A window lock switch is included on four-door model master control switches. When open, this switch limits opening and closing of all windows to the master control switch. It is included as a safety device to prevent children from opening door windows without the driver knowing.

The power windows on many vehicles are controlled by the driver's door module or DDM. The DDM is part of the CAN network in the vehicle. The window switches are only used as inputs to the module and do not carry the heavy current to and from the window motor, thus reducing the size and amount of wiring in the vehicle.

Power Seats

Power seats allow the driver or passenger to adjust the seat to the most comfortable position.

The major components of the system are the seat control and the motors. In the four-way system the whole seat moves up or down, or forward and rearward. In the six-way system, the same movements are included plus the ability to adjust the height of the seat front or the seat rear. Two motors are usually employed to make the adjustments on the four-way systems, while three are used on a six-way system.

Power Mirror System

The power mirror system allows the driver to control both the left-hand and right-hand outside rear-view mirrors from one switch. The major components in the system are the joystick control switch and a dual motor drive assembly located in each mirror assembly.

Horns

Most horn systems are controlled by relays. When the horn button, ring, or padded unit is depressed, electricity flows from the battery through a horn lead, into an electromagnetic coil in the horn relay to the ground. A small flow of electric current through the coil energizes the electromagnet, pulling a movable arm. Electrical contacts on the arm touch, closing the primary circuit and causing the horn to sound.

Cruise (Speed) Control System

Cruise or speed control systems are designed to allow the driver to maintain a constant speed without having to apply continual foot pressure on the accelerator pedal. Selected cruise speeds are easily maintained and speed can be easily changed. Several override systems also allow the vehicle to be accelerated, slowed, or stopped. Because of the constant changes and improvements in technology, each cruise control system may be considerably different from others. The cruise control switch usually has several positions, including off/on, resume, and engage buttons. With the advent of throttle actuator control (TAC) (also called throttle by wire) the cruise control servo and linkage have been eliminated. The throttle position is maintained by the ECM

Cruise control can be obtained by using mechanical or electronic components, depending upon the manufacturer. Some rely on electronic controls to regulate the mechanical system. Vehicle with electronic throttle control rely solely on electronics. These systems offer precise speed control. The electronic control module receives several inputs that help determine the operation of the servo. These inputs include a brake release switch (clutch release switch), a vehicle speed sensor, and controls on the steering wheel (signal to control the cruise control).

Hybrid Systems

A hybrid electric vehicle (HEV) uses one or more electric motors and an engine to propel the vehicle. Depending on the design of the system, the engine may move the vehicle by itself, assist the electric motor while it is moving the vehicle, or it may drive a generator to charge the vehicle's batteries. The electric motor may power the vehicle by itself or assist the engine while it is propelling the vehicle. Many hybrids rely exclusively on the electric motor(s) during slow speed operation, the engine at higher speeds, and both during some certain driving conditions. Complex electronic controls monitor the operation of the vehicle, based on the current operating conditions; electronics control the engine, electric motor, and generator.

A hybrid's electric motor is powered by high-voltage batteries, which are recharged by a generator driven by the engine and through regenerative braking. Regenerative braking is the process by which a vehicle's kinetic energy can be captured while it is decelerating and braking. The electric drive motors become generators driven by the vehicle's wheels. These generators take the kinetic energy, or the energy of the moving vehicle, and change it into energy that charges the batteries. The magnetic forces inside the generator cause the drive wheels to slow down. A conventional brake system brings the vehicle to a safe stop.

The engines used in hybrids are specially designed for the vehicle and electric assist. Therefore, they can operate more efficiently; resulting in very good fuel economy and very low tailpipe emissions. HEVs can provide the same performance, if not better, as a comparable vehicle equipped with a larger engine.

There are primarily two types of hybrids: the parallel and the series designs. A parallel hybrid electric vehicle uses either the electric motor or the gas engine to propel the vehicle, or both.

The engine in a true series hybrid electric vehicle is used only to drive the generator that keeps the batteries charged. The vehicle is powered only by the electric motor(s). Most current HEVs are considered as having a series/parallel configuration because they have the features of both designs.

Although most current hybrids are focused on fuel economy, the same ideas can be used to create high-performance vehicles. Hybrid technology is also influencing off-the-road performance. By using individual motors at the front and rear drive axles, additional power can be applied to certain drive wheels, when needed.

SAFETY

In an automotive repair shop, there is great potential for serious accidents, simply because of the nature of the business and the equipment used. When people are careless, the automotive repair industry can be one of the most dangerous occupations. But, the chances of your being injured while working on a car are close to nil if you learn to work safely and use common sense. Safety is the responsibility of everyone in the shop.

Personal Protection

Some procedures, such as grinding, result in tiny particles of metal and dust that are thrown off at very high speeds. These metal and dirt particles can easily get into your eyes, causing scratches or cuts on your eyeball. Pressurized gases and liquids escaping a ruptured hose or hose-fitting can spray a great distance. If these chemicals get into your eyes, they can cause blindness. Dirt and sharp bits of corroded metal can easily fall down into your eyes while you are working under a vehicle.

Eye protection should be worn whenever you are exposed to these risks. To be safe, you should wear safety glasses whenever you are working in the shop. Some procedures may require that you wear other eye protection in addition to safety glasses. For example, when cleaning parts with a pressurized spray, you should wear a face shield. The face shield not only gives added protection to your eyes but also protects the rest of your face.

If chemicals such as battery acid, fuel, or solvents get into your eyes, flush them continuously with clean water. Have someone call a doctor and get medical help immediately.

Your clothing should be well fitted and comfortable but made of strong material. Loose, baggy clothing can easily get caught in moving parts and machinery. Some technicians prefer to wear coveralls or shop coats to protect their personal clothing. Your work clothing should offer you some protection but should not restrict your movement.

Long hair and loose, hanging jewelry can create the same type of hazard as loose-fitting clothing. They can get caught in moving engine parts and machinery. If you have long hair, tie it back or tuck it under a cap.

Never wear rings, watches, bracelets, and neck chains. These can easily get caught in moving parts and cause serious injury.

Always wear shoes or boots of leather or similar material with non-slip soles. Steel-tipped safety shoes can give added protection to your feet. Jogging or basketball shoes, street shoes, and sandals are inappropriate in the shop.

Good hand protection is often overlooked. A scrape, cut, or burn can limit your effectiveness at work for many days. A well-fitted pair of heavy work gloves should be worn during operations such as grinding and welding or when handling hot components. Always wear approved rubber gloves when handling strong and dangerous caustic chemicals.

Many technicians wear thin, surgical-type latex or vinyl gloves whenever they are working on vehicles. These offer little protection against cuts but do offer protection against disease and grease buildup under and around your fingernails. These gloves are comfortable and are quite inexpensive.

Accidents can be prevented simply by the way you act. The following are some guidelines to follow while working in a shop. This list does not include everything you should or shouldn't do; it merely presents some things to think about.

- Never smoke while working on a vehicle or while working with any machine in the shop.
- Playing around is not fun when it sends someone to the hospital.
- To prevent serious burns, keep your skin away from hot metal parts such as the radiator, exhaust manifold, tailpipe, catalytic converter, and muffler.
- Always disconnect electric engine cooling fans when working around the radiator.

Many of these will turn on without warning and can easily chop off a finger or hand. Make sure you reconnect the fan after you have completed your repairs.

■ When working with a hydraulic press, make sure the pressure is applied in a safe manner. It is generally wise to stand to the side when operating the press.

■ Properly store all parts and tools by putting them away in a place where people will not trip over them. This practice not only cuts down on injuries, it also reduces time wasted looking for a misplaced part or tool.

Work Area Safety

Your entire work area should be kept clean and safe. Any oil, coolant, or grease on the floor can make it slippery. To clean up oil, use commercial oil absorbent. Keep all water off the floor. Water makes smooth floors slippery and is dangerous as a conductor of electricity. Aisles and walkways should be kept clean and wide enough to allow easy movement. Make sure the work areas around machines are large enough to allow the machinery to be operated safely.

Gasoline is a highly flammable volatile liquid. Something that is flammable catches fire and burns easily. A volatile liquid is one that vaporizes very quickly. Flammable volatile liquids are potential firebombs. Always keep gasoline or diesel fuel in an approved safety can and never use gasoline to clean your hands or tools.

Handle all solvents (or any liquids) with care to avoid spillage. Keep all solvent containers closed, except when pouring. Proper ventilation is very important in areas where volatile solvents and chemicals are used. Solvent and other combustible materials must be stored in approved and designated storage cabinets or rooms with adequate ventilation. Never light matches or smoke near flammable solvents and chemicals, including battery acids.

Oily rags should also be stored in an approved metal container. When these oily, greasy, or paint-soaked rags are left lying about or are not stored properly, they can cause spontaneous combustion. Spontaneous combustion results in a fire that starts by itself, without a match.

Disconnecting the vehicle's battery before working on the electrical system, or before welding, can prevent fires caused by a vehicle's electrical system. To disconnect the battery, remove the negative or ground cable from the battery and position it away from the battery.

Working on a hybrid vehicle takes some unique precautions, because of the increased high voltage used by the high-voltage system. The current and voltage are high enough in the system to KILL you instantly. Never ever work on a hybrid vehicle without factory training. This book does not include the necessary training to work on a hybrid vehicle. The information contained is for informational purposes only! More on hybrid vehicles is included below.

Know where all of the shop's fire extinguishers are located. Fire extinguishers are clearly labeled as to what type they are and what types of fire they should be used on. Make sure you use the correct type of extinguisher for the type of fire you are dealing with. A multipurpose dry chemical fire extinguisher will put out ordinary combustibles, flammable liquids, and electrical fires. Never put water on a gasoline fire because it will just cause the fire to spread. The proper fire extinguisher will smother the flames.

During a fire, never open doors or windows unless it is absolutely necessary; the extra draft will only make the fire worse. Make sure the fire department is contacted before or during your attempt to extinguish a fire.

Conventional Battery Safety

The potential dangers caused by the sulfuric acid in the electrolyte and the explosive gases generated during battery charging require that battery service and troubleshooting are conducted under absolutely safe working conditions. Always wear safety glasses or goggles when working with batteries no matter how small the job.

Sulfuric acid can also cause severe skin burns. If electrolyte contacts your skin or eyes, flush the area with water for several minutes. When eye contact occurs, force your eyelid open. Always have a bottle of neutralizing eyewash on hand and flush the affected areas with it. Do not rub your eyes or skin. Seek prompt medical attention if electrolyte contacts your skin or eyes. Call a doctor immediately.

When a battery is charging or discharging, it gives off quantities of highly explosive hydrogen gas. Some hydrogen gas is present in the battery at all times. Any flame or spark can ignite this gas, causing the battery to explode violently, propelling the vent caps at a high velocity and spraying acid over a wide area. To prevent this dangerous situation take these precautions:

- Never smoke near the top of a battery and never use a lighter or match as a flashlight.
- Remove wristwatches and rings before servicing any part of the electrical system. This helps to prevent the possibility of electrical arcing and burns.
- Even sealed, maintenance-free batteries have vents and can produce dangerous quantities of hydrogen if severely overcharged.
- When removing a battery from a vehicle, always disconnect the battery ground cable first. When installing a battery, connect the ground cable last.
- Always disconnect the battery's ground cable when working on the electrical system or engine. This prevents sparks from short circuits and prevents accidental starting of the engine.
- Always operate charging equipment in well-ventilated areas. A battery that has been overworked should be allowed to cool down; let air circulate around it before attempting to jump start the vehicle. Most batteries have flame arresters in the caps to help prevent explosions, so make sure that the caps are tightly in place.
- Never connect or disconnect charger leads when the charger is turned on. This generates a dangerous spark.
- Never lay metal tools or other objects on the battery, because a short circuit across the terminals can result.
- Always disconnect the battery ground cable before fast-charging the battery on the vehicle. Improper connection of charger cables to the battery can reverse the current flow and damage the AC generator.
- Never attempt to use a fast charger as a boost to start the engine.
- As a battery gets closer to being fully discharged, the acidity of the electrolyte is reduced, and the electrolyte starts to behave more like pure water. A dead

battery may freeze at temperatures near 0°F. Never try to charge a battery that has ice in the cells. Passing current through a frozen battery can cause it to rupture or explode. If ice or slush is visible or the electrolyte level cannot be seen, allow the battery to thaw at room temperature before servicing. Do not take chances with sealed batteries. If there is any doubt, allow them to warm to room temperature before servicing.

- As batteries get old, especially in warm climates and especially if they have lead-calcium cells, the grids start to grow. The chemistry is rather involved, but the point is that plates can grow to the point where they touch, producing a shorted cell.
- Always use a battery carrier or lifting strap to make moving and handling batteries easier and safer.
- Acid from the battery damages a vehicle's paint and metal surfaces and harms shop equipment. Neutralize any electrolyte spills during servicing.

Air Bag Safety

When service is performed on any air bag system component, always disconnect the negative battery cable, isolate the cable end, and wait for the amount of time specified by the vehicle manufacturer before proceeding with the necessary diagnosis or service. The average waiting period is two minutes, but some vehicle manufacturers specify up to ten minutes. Failure to observe this precaution may cause accidental air bag deployment and personal injury.

Replacement air bag system parts must have the same part number as the original part. Replacement parts of lesser or questionable quality must not be used. Improper or inferior components may result in inappropriate air bag deployment and injury to the vehicle occupants.

Do not strike or jar a sensor or an air bag system diagnostic monitor (ASDM). This may cause air bag deployment or make the sensor inoperative. Accidental air bag deployment may cause personal injury, and an inoperative sensor may result in air bag deployment failure, causing personal injury to vehicle occupants.

All sensors and mounting brackets must be properly torqued to ensure correct sensor operation before an air bag system is powered up. If sensor fasteners do not have the proper torque, improper air bag deployment may result in injury to vehicle occupants.

When working on the electrical system on an air-bag-equipped vehicle, use only the vehicle manufacturer's recommended tools and service procedures. The use of improper tools or service procedures may cause accidental air bag deployment and personal injury. For example, do not use 12V or self-powered test lights when servicing the electrical system on an air-bag-equipped vehicle.

Tool and Equipment Safety

Careless use of simple hand tools such as wrenches, screwdrivers, and hammers causes many shop accidents that could be prevented. Keep all hand tools grease-free and in good condition. Tools that slip can cause cuts and bruises. If a tool slips and falls into a moving part, it can fly out and cause serious injury.

Use the proper tool for the job. Make sure the tool is of professional quality. Using poorly made tools or the wrong tools can damage parts or the tool itself, or could cause injury. Never use broken or damaged tools.

Safety around power tools is very important. Serious injury can result from carelessness. Always wear safety glasses when using power tools. If the tool is electrically powered, make sure it is properly grounded. Before using it, check the wiring for cracks in the insulation, as well as for bare wires. Also, when using electrical power tools, never stand on a wet or damp floor. Never leave a running power tool unattended.

When using compressed air, safety glasses and/or a face shield should be worn. Particles of dirt and pieces of metal, blown by the high-pressure air, can penetrate your skin or get into your eyes.

Always be careful when raising a vehicle on a lift or a hoist. Adapters and hoist plates must be positioned correctly to prevent damage to the underbody of the vehicle. There are specific lift points that allow the weight of the vehicle to be evenly supported by the adapters or hoist plates. The correct lift points can be found in the vehicle's service information. Before operating any lift or hoist, carefully read the operating manual and follow the operating instructions.

Once you feel the lift supports are properly positioned under the vehicle, raise the lift until the supports contact the vehicle. Then, check the supports to make sure they are in full contact with the vehicle. Shake the vehicle to make sure it is securely balanced on the lift, and then raise the lift to the desired working height. Before working under a car, make sure the lift's locking devices are engaged.

A vehicle can be raised off the ground by a hydraulic jack. The jack's lifting pad must be positioned under an area of the vehicle's frame or at one of the manufacturer's recommended lift points. Never place the pad under the floor pan or under steering and suspension components because they are easily damaged by the weight of the vehicle. Always position the jack so the wheels of the vehicle can roll as the vehicle is being raised.

Safety stands, also called jack stands, should be placed under a sturdy chassis member, such as the frame or axle housing, to support the vehicle after it has been raised by a jack. Once the safety stands are in position, the hydraulic pressure in the jack should be slowly released until the weight of the vehicle is on the stands. Never move under a vehicle when it is supported only by a hydraulic jack. Rest the vehicle on the safety stands before moving under the vehicle.

Parts cleaning is a necessary step in most repair procedures. Always wear the appropriate protection when using chemical, abrasive, and thermal cleaners.

Working Safely on High-Voltage Systems

Electric drive vehicles (battery operated, hybrid, and fuel cell electric vehicles) have high-voltage electrical systems (from 42-volts to 650-volts). These high-voltages can kill you! Fortunately, most high-voltage circuits are identifiable by size and color. The cables have thicker insulation and are typically colored orange. The connectors are also colored orange. On some vehicles, the high-voltage cables are enclosed in an orange shielding or casing, again the orange indicates high voltage. In addition, the high-voltage battery pack and most high-voltage components have "High Voltage" caution labels. Be careful not to touch these wires and parts.

Wear insulating gloves, commonly called "lineman's gloves", when working on or around the high-voltage system. These gloves must be class "0" rubber insulating gloves, rated at 1000-volts. Also, to protect the integrity of the insulating gloves, as well as you, wear leather gloves over the insulating gloves while doing a service.

Make sure they have no tears, holes or cracks and that they are dry. Electrons are very small and can enter through the smallest of holes in your gloves. The integrity of the gloves should be checked before using them. To check the condition of the gloves, blow enough air into each one so they balloon out. Then fold the open end over to seal the air in. Continue to slowly fold that end of the glove toward the fingers. This will compress the air. If the glove continues to balloon as the air is compressed, it has no leaks. If any air leaks out, the glove should be discarded. All gloves, new and old, should be checked before they are used.

There are other safety precautions that should always be adhered to when working on an electric drive vehicle:

- Always adhere to the safety guidelines given by the vehicle's manufacture.
- Obtain the necessary training before working on these vehicles.
- Be sure to perform each repair operation following the test procedures defined by the manufacturer.
- Disable or disconnect the high-voltage system before performing services to those systems. Do this according to the procedures given by the manufacturer.
- Anytime the engine is running in a hybrid vehicle, the generator is producing high-voltage and care must be taken to prevent being shocked.
- Before doing any service to an electric drive vehicle, make sure the power to the electric motor is disconnected or disabled.
- Systems may have a high-voltage capacitor that must be discharged after the high-voltage system has been isolated. Make sure to wait the prescribed amount of time (normally about 10 minutes) before working on or around the high-voltage system.
- After removing a high-voltage cable, cover the terminal with vinyl electrical tape.

- Always use insulated tools.
- Alert other technicians that you are working on the high-voltage systems with a warning sign such as "High Voltage Work: Do Not Touch."
- Always install the correct type of circuit protection device into a high-voltage circuit.
- Many electric motors have a strong permanent magnet in them; individuals with a pacemaker should not handle these parts.
- When an electric drive vehicle needs to be towed into the shop for repairs, make sure it is not towed on its drive wheels. Doing this will drive the generator(s), which can overcharge the batteries and cause them to explode. Always tow these vehicles with the drive wheels off the ground or move them on a flat bed.

Vehicle Operation

When the customer brings a vehicle in for service, certain driving rules should be followed to ensure your safety and the safety of those working around you. For example, before moving a car into the shop, buckle your safety belt. Make sure no one is near, the way is clear, and there are no tools or parts under the car before you start the engine. Check the brakes before putting the vehicle in gear. Then, drive slowly and carefully in and around the shop.

If the engine must be running while you are working on the car, block the wheels to prevent the car from moving. Place the transmission into park for automatic transmissions or into neutral for manual transmissions. Set the parking (emergency) brake. Never stand directly in front of or behind a running vehicle.

Run the engine only in a well-ventilated area to avoid the danger of poisonous carbon monoxide (CO) in the engine exhaust. CO is an odorless but deadly gas. Most shops have an exhaust ventilation system and you should always use it. Connect the hose from the vehicle's tailpipe to the intake for the vent system. Make sure the vent system is turned on before running the engine. If the work area does not have an exhaust venting system, use a hose to direct the exhaust out of the building.

HAZARDOUS MATERIALS AND WASTES

A typical shop contains many potential health hazards for those working in it. These hazards can cause injury, sickness, health impairments, discomfort, and even death. Here is a short list of the different classes of hazards:

- Chemical hazards are caused by high concentrations of vapors, gases, or solids in the form of dust.
- Hazardous wastes are those substances that result from performing a service.
- Physical hazards include excessive noise, vibration, pressures, and temperatures.
- Ergonomic hazards are conditions that impede normal and/or proper body position and motion.

There are many government agencies charged with ensuring safe work environments for all workers. These include the Occupational Safety and Health Administration (OSHA), Mine Safety and Health Administration (MSHA), and National Institute for Occupational Safety and Health (NIOSH). These, as well as state and local governments, have instituted regulations that must be understood and followed. Everyone in a shop has the responsibility for adhering to these regulations. An important part of a safe work environment is the employees' knowledge of potential hazards. Right-to-know laws concerning all chemicals protect every employee in the shop. The general intent of right-to-know laws is to ensure that employers provide their employees with a safe working place as far as hazardous materials are concerned.

All employees must be trained about their rights under the legislation, the nature of the hazardous chemicals in their workplace, and the contents of the labels on the chemicals. All of the information about each chemical must be posted on material safety data sheets (MSDS) and must be accessible. The manufacturer of the chemical must give these sheets to its customers, if they are requested to do so. They detail the chemical composition and precautionary information for all products that can present a health or safety hazard.

Employees must become familiar with the general uses, protective equipment, accident or spill procedures, and any other information regarding the safe handling of the hazardous material. This training must be given to employees annually and provided to new employees as part of their job orientation.

A hazardous material must be properly labeled, indicating what health, fire, or reactivity hazard it poses and what protective equipment is necessary when handling each chemical. The manufacturer of the hazardous materials must provide all warnings and precautionary information, which must be read and understood by the user before use. A list of all hazardous materials used in the shop must be posted for the employees to see.

Shops must maintain documentation on the hazardous chemicals in the workplace, proof of training programs, records of accidents or spill incidents, satisfaction of employee requests for specific chemical information via the MSDS, and a general right-to-know compliance procedure manual utilized within the shop.

When handling any hazardous materials or hazardous waste, make sure you follow the required procedures for handling such material. Also wear the proper safety equipment listed on the MSDS. This includes the use of approved respirator equipment.

Some of the common hazardous materials that automotive technicians use are: cleaning chemicals, fuels (gasoline and diesel), paints and thinners, battery electrolyte (acid), used engine oil, refrigerants, and engine coolant (anti-freeze).

Many repair and service procedures generate what are known as hazardous wastes. Dirty solvents and cleaners are good examples of hazardous wastes. Something is classified as a hazardous waste if it is on the EPA list of known harmful materials or has one or more of the following characteristics.

- *Ignitability*. If it is a liquid with a flash point below 140°F or a solid that can spontaneously ignite.
- *Corrosivity*. If it dissolves metals and other materials or burns the skin.
- *Reactivity*. Any material that reacts violently with water or other materials or releases cyanide gas, hydrogen sulfide gas, or similar gases when exposed to low pH acid solutions. This also includes material that generates toxic mists, fumes, vapors, and flammable gases.

■ *Toxicity*. Materials that leach one or more of eight heavy metals in concentrations greater than 100 times primary drinking water standard concentrations.

Complete EPA lists of hazardous wastes can be found in the Code of Federal Regulations. It should be noted that no material is considered hazardous waste until the shop is finished using it and is ready to dispose of it.

The following list covers the recommended procedure for dealing with some of the common hazardous wastes. Always follow these and any other mandated procedures.

Oil Recycle oil. Set up equipment, such as a drip table or screen table with a used oil collection bucket, to collect oils dripping off parts. Place drip pans underneath vehicles that are leaking fluids onto the storage area. Do not mix other wastes with used oil, except as allowed by your recycler. Used oil generated by a shop (and/or oil received from household "do-it-yourself" generators) may be burned on site in a commercial space heater. Also, used oil may be burned for energy recovery. Contact state and local authorities to determine requirements and to obtain necessary permits.

Oil filters Drain for at least 24 hours, crush, and recycle used oil filters.

Batteries Recycle batteries by sending them to a reclaimer or back to the distributor. Keeping shipping receipts can demonstrate that you have done the recycling. Store batteries in a watertight, acid-resistant container. Inspect batteries for cracks and leaks when they come in. Treat a dropped battery as if it were cracked. Acid residue is hazardous because it is corrosive and may contain lead and other toxic substances. Neutralize spilled acid, by using baking soda or lime, and dispose of it as a hazardous material.

Metal residue from machining Collect metal filings when machining metal parts. Keep them separate and recycle if possible. Prevent metal filings from falling into a storm sewer drain.

Refrigerants Recover and/or recycle refrigerants during the servicing and disposal of motor vehicle air conditioners and refrigeration equipment. It is not allowable to knowingly vent refrigerants to the atmosphere.

Recovering and/or recycling during servicing must be performed by an EPA-certified technician using certified equipment and following specified procedures.

Solvents Replace hazardous chemicals with less toxic alternatives that perform equally. For example, substitute water-based cleaning solvents for petroleum-based solvent degreasers. To reduce the amount of solvent used when cleaning parts, use a two-stage process: dirty solvent followed by fresh solvent. Hire a hazardous waste management service to clean and recycle solvents. (Some spent solvents must be disposed of as hazardous waste, unless recycled properly.) Store solvents in closed containers to prevent evaporation. Evaporation of solvents contributes to ozone depletion and smog formation. In addition, the residue from evaporation must be treated as a hazardous waste. Properly label spent solvents and store on drip pans or in diked areas and only with compatible materials.

Containers Cap, label, cover, and properly store aboveground outdoor liquid containers and small tanks within a diked area and on a paved impermeable surface to prevent spills from running into surface or ground water.

Other solids Store materials such as scrap metal, old machine parts, and worn tires under a roof or tarpaulin to protect them from the elements and to prevent the possibility of creating contaminated runoff. Consider recycling tires by retreading them.

Liquid recycling Collect and recycle coolants from radiators. Store transmission fluids, brake fluids, and solvents containing chlorinated hydrocarbons separately, and recycle or dispose of them properly.

Shop towels or rags Keep waste towels in a closed container marked "contaminated shop towels only." To reduce costs and liabilities associated with disposal of used towels (which can be classified as hazardous wastes); investigate using a laundry service that is able to treat the wastewater generated from cleaning the towels.

Waste storage Always keep hazardous waste separate, properly labeled, and sealed in the recommended containers. The storage area should be covered and may need to be fenced and locked if vandalism could be a problem. Select a licensed hazardous waste hauler after seeking recommendations and reviewing the firm's permits and authorizations.

ELECTRICAL AND ELECTRONIC SYSTEMS TOOLS AND EQUIPMENT

Many different tools and many kinds of testing and measuring equipment are used to service electrical and electronic systems. NATEF has identified many of these and has said an Electrical/Electronic technician must know what they are and how and when to use them. The tools and equipment listed by NATEF are covered in the following discussion. Also included are the tools and equipment you will use while completing the job sheets. Although you need to be more than familiar with and will be using common hand tools, they are not part of this discussion. You should already know what they are and how to use and care for them.

Circuit Tester

Circuit testers are used to identify short and open circuits in any electrical circuit. Low-voltage testers are used to troubleshoot 6- to 12-volt circuits. A circuit tester, commonly called a test light, looks like a stubby ice pick. Its handle is transparent and contains a light bulb. A probe extends from one end of the handle and a ground clip and wire from the other end. When the ground clip is attached to a good ground and the probe touched to a live connector, the bulb in the handle will light up. If the bulb does not light, voltage is not available at the connector.

WARNING: *Do not use a conventional 12-V test light to diagnose components and wires in electronic systems. The current draw of these test lights may damage computers and system components. High-impedance test lights are available for diagnosing electronic systems.*

A self-powered test light is called a continuity tester. It is used on non-powered circuits. It looks like a regular test light, except that it has a small internal battery. When the ground clip is attached to the negative side of a component and the probe touched to the positive side, the lamp will light if there is continuity in the circuit. If an open circuit exists, the lamp will not light. Do not use any type of test light or circuit tester to diagnose automotive air bag systems.

Voltmeter

A voltmeter has two leads: a red positive lead and a black negative lead. The red lead should be connected to the positive side of the circuit or component. The black should be connected to ground or to the negative side of the component. Voltmeters should be connected across the circuit being tested.

The voltmeter measures the voltage available at any point in an electrical system. A voltmeter can also be used to test voltage drop across an electrical circuit, component, switch, or connector. A voltmeter can also be used to check for proper circuit grounding.

Ohmmeter

An ohmmeter measures the resistance to current flow in a circuit. In contrast to the voltmeter, which uses the voltage available in the circuit, the ohmmeter is battery powered. The circuit being tested must be open. If the power is on in the circuit, the ohmmeter will be damaged.

The two leads of the ohmmeter are placed across or in parallel with the circuit or component being tested. The red lead is placed on the positive side of the circuit and the black lead is placed on the negative side of the circuit. The meter sends current through the component and determines the amount of resistance based on the voltage dropped across the load. The scale of an ohmmeter reads from zero to infinity. A zero reading means there is no resistance in the circuit and may indicate a short in a component that should show a specific resistance. An infinite reading indicates a number higher than the meter can measure. This usually is an indication of an open circuit.

Ohmmeters are also used to trace and check wires or cables. Assume that one wire of a four-wire cable is to be found. Connect one probe of the ohmmeter to the known wire at one end of the cable and touch the other probe to each wire at the other end of the cable. Any evidence of resistance, such as meter needle deflection, indicates the correct wire. Using this same method, you can check a suspected defective wire. If resistance is shown on the meter, the wire is sound.

If no resistance is measured, the wire is defective (open). If the wire is okay, continue checking by connecting the probe to other leads. Any indication of resistance indicates that the wire is shorted to one of the other wires and that the harness is defective.

Ammeter

An ammeter measures current flow in a circuit. The ammeter must be placed into the circuit or in series with the circuit being tested. Normally, this requires disconnecting a wire or connector from a component and connecting the ammeter between the wire or connector and the component. The red lead of the ammeter should always be connected to the side of the connector closest to the positive side of the battery and the black lead should be connected to the other side.

It is much easier to test current using an ammeter with an inductive pickup. The pickup clamps around the wire or cable being tested. These ammeters measure amperage based on the magnetic field created by the current flowing through the wire. This type of pickup eliminates the need to separate the circuit to insert the meter.

Because ammeters are built with very low internal resistance, connecting them in series does not add any appreciable resistance to the circuit. Therefore, an accurate measurement of the current flow can be taken.

Volt/Ampere Tester

A volt/ampere tester (VAT) is used to test batteries, starting systems, and charging systems (Figure 2). The tester contains a voltmeter, ammeter, and carbon pile. The carbon pile is a variable resistor. A knob on the tester allows the technician to vary the resistance of the pile. When the tester is attached to the battery, the carbon pile will draw current out of the battery. The ammeter will read the amount of current draw. When testing a battery, the resistance of the carbon pile must be adjusted to match the ratings of the battery.

Conductance Tester

A conductance, or capacitance, tester is another battery tester. Conductance is a measurement of the battery's ability to produce current. To measure conductance, the tester creates an AC

Figure 2 A VAT used to test batteries and starting and charging systems.

voltage of a known frequency and amplitude that is sent through the battery, and then measures a portion of the AC current response. The degree of conductance is based on the plate surface that is available in the battery, which determines how much power the battery can supply.

As a battery ages, the plate surface can sulfate or shed active material, which adversely affects its ability to perform. In addition, conductance can be used to detect cell defects, shorts, and open circuits, which will reduce the ability of the battery to deliver current. Using conductance, the testers are able to provide an accurate assessment of the battery's condition.

To use this tester, connect the two test leads to the positive and negative posts of the cell or battery being tested, a measurement is taken in a matter of seconds. A conductance measurement is displayed in Mhos or Siemens, sometimes abbreviated with a "G."

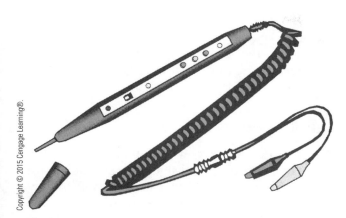

Copyright © 2015 Cengage Learning®.

Figure 3 A logic probe.

Logic Probes

In some circuits, pulsed or digital signals pass through the wires. These "on-off" digital signals either carry information or provide power to drive a component. Many sensors used in a computer-control circuit send digital information back to the computer. To check the continuity of the wires that carry digital signals, a logic probe can be used.

A logic probe (Figure 3) has three different colored LEDs. A red LED lights when there is high voltage at the point being probed. A green LED lights to indicate low voltage. And a yellow LED indicates the presence of a voltage pulse. The logic probe is powered by the circuit and reflects only the activity at the point being probed. When the probe's test leads are attached to a circuit, the LEDs display the activity.

If a digital signal is present, the yellow LED will turn on. When there is no signal, the LED is off. If voltage is present, the red or green LEDs will light, depending on the amount of voltage. When there is a digital signal and the voltage cycles from low to high, the yellow LED will be lit and the red and green LEDs will cycle, indicating a change in the voltage.

DMMs

It's not necessary for a technician to own separate meters to measure volts, ohms, and amps; a multimeter can be used instead. Top-of-the-line multimeters are multifunctional. Most test volts, ohms, and amperes in both DC and AC. Usually there are several test ranges provided for each of these functions. In addition to these basic electrical tests, multimeters also test engine rpm, duty cycle, pulse width, diode condition, frequency,

and even temperature. The technician selects the desired test range by turning a control knob on the front of the meter.

Multimeters are available with either analog or digital displays, but the most commonly used multimeter is the digital volt/ohmmeter (DVOM), which is often referred to as a digital multimeter (DMM). There are several drawbacks to using analog-type meters for testing electronic control systems. Many electronic components require very precise test results. Digital meters can measure volts, ohms, or amperes in tenths and hundredths. Another problem with analog meters is their low internal resistance (input impedance). The low input impedance allows too much current to flow through circuits and should not be used on delicate electronic devices.

Digital meters, on the other hand, have high input impedance, usually at least 10 megohms (10 million ohms). Metered voltage for resistance tests is well below 5 volts, reducing the risk of damage to sensitive components and delicate computer circuits. A high-impedance digital multimeter must be used to test the voltage of some components and systems such as an oxygen (O_2) sensor circuit. If a low-impedance analog meter is used in this type of circuit, the current flow through the meter is high enough to damage the sensor.

DMMs have either an "auto range" feature, in which the appropriate scale is automatically selected by the meter, or they must be set to a particular range. In either case, you should be familiar with the ranges and the different settings available on the meter you are using. To designate particular ranges and readings, meters display a prefix before the reading or range. If the meter has a setting for mAmps, that means the readings will be given in milli-amps or 1/1000th of an amp. Ohmmeter scales are expressed as a multiple of tens or use the prefix K or M. K stands for Kilo or 1000. A reading of 10K ohms equals 10,000 ohms. An M stands for Mega or 1,000,000. A reading of 10M ohms equals 10,000,000 ohms. When using a meter with an auto range, make sure you note the range being used by the meter. There is a big difference between 10 ohms and 10,000,000 ohms.

After the test range has been selected, the meter is connected to the circuit in the same way as if it were an individual meter.

When using the ohmmeter function, the DMM will show a zero or close to zero when

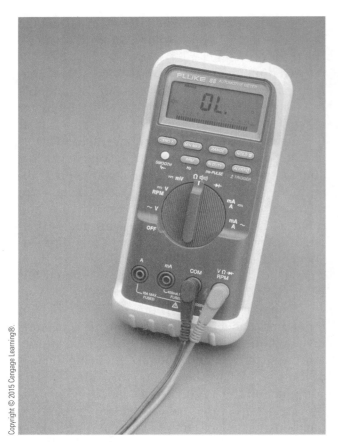

Figure 4 The infinite reading on a DMM.

there is good continuity. If the continuity is very poor, the meter will display an infinite reading. This reading is usually shown as a blinking "1.000," a blinking "1," or an "OL" (Figure 4). Before taking any measurement, calibrate the meter. This is done by holding the two leads together and adjusting the meter reading to zero. Not all meters need to be calibrated; some digital meters automatically calibrate when a scale is selected. On meters that require calibration, it is recommended that the meter be zeroed after changing scales.

Multimeters may also have the ability to measure duty cycle, pulse width, and frequency. All of these represent voltage pulses caused by the turning on and off of a circuit or the increase and decrease of voltage in a circuit. Duty cycle is a measurement of the amount of time something is on compared to the time of one cycle and is measured in a percentage.

Pulse width is similar to duty cycle except that it is the exact time something is turned on and is measured in milliseconds. When measuring duty cycle, you are looking at the amount of time something is on during one cycle.

The number of cycles that occur in one second is called the frequency. The higher the frequency, the more cycles occur in a second. Frequencies are measured in Hertz. One Hertz is equal to one cycle per second.

Lab Scopes

An oscilloscope is a visual voltmeter. An oscilloscope converts electrical signals to a visual image representing voltage changes over a specific period of time. This information is displayed in the form of a continuous voltage line called a waveform pattern or trace.

An oscilloscope screen is a cathode ray tube (CRT), which is very similar to the picture tube in a television set. High voltage from an internal source is supplied to an electron gun in the back of the CRT when the oscilloscope is turned on. This electron gun emits a continual beam of electrons against the front of the CRT. The external leads on the oscilloscope are connected to deflection plates above and below and on each side of the electron beam. When a voltage signal is supplied from the external leads to the deflection plates, the electron beam is distorted and strikes the front of the screen in a different location to indicate the voltage signal from the external leads.

An upward movement of the voltage trace on an oscilloscope screen indicates an increase in voltage, and a downward movement of this trace represents a decrease in voltage. As the voltage trace moves across an oscilloscope screen, it represents a specific length of time.

The size and clarity of the displayed waveform is dependent on the voltage scale and the time reference selected. Most scopes are equipped with controls that allow voltage and time interval selection. It is important, when choosing the scales, to remember that a scope displays voltage over time.

Dual-trace oscilloscopes can display two different waveform patterns at the same time (Figure 5). This makes cause and effect analysis easier.

With a scope, precise measurement is possible. A scope will display any change in voltage as it occurs. This is especially important for diagnosing intermittent problems.

The screen of a lab scope is divided into small divisions of time and voltage. Time is represented by the horizontal movement of the waveform. Voltage

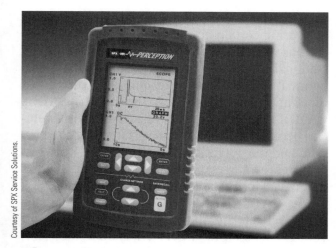

Figure 5 A hand-held dual-trace lab scope.

is measured with the vertical position of the waveform. Since the scope displays voltage over time, the waveform moves from the left (the beginning of measured time) to the right (the end of measured time). The value of the divisions can be adjusted to improve the view of the voltage waveform.

Since a scope displays actual voltage, it will display any electrical noise or disturbances that accompany the voltage signal. Noise is primarily caused by radio frequency interference (RFI), which may come from the ignition system. RFI is an unwanted voltage signal that rides on a signal.

This noise can cause intermittent problems with unpredictable results. The noise causes slight increases and decreases in the voltage. When a computer receives a voltage signal with noise, it will try to react to the minute changes. As a result, the computer responds to the noise rather than the voltage signal.

Graphing Multimeter

Graphing digital multimeters display readings over time, similar to a lab scope. The graph displays the minimum and maximum readings on a graph, as well as displaying the current reading. By observing the graph, a technician can see any undesirable changes during the transition from a low reading to a high reading, or vice versa. These glitches are some of the more difficult problems to identify without a graphing meter or a lab scope.

Scan Tools

The introduction of computer-controlled systems brought with it the need for tools capable of troubleshooting electronic control systems. There are a variety of scan tools available today that do just that. A scan tool is a microprocessor designed to communicate with the vehicle's computer. Connected to the computer through diagnostic connector, a scan tool can access trouble codes, run tests to check system operations, and monitor the activity of the system.

Scan tools are capable of testing many onboard computer systems, such as climate controls, transmission controls, engine computers, antilock brake computers, air bag computers, and suspension computers, depending on the year and make of the vehicle and the type of scan tester. In many cases, the technician must select the computer system to be tested with the scanner after it has been connected to the vehicle.

With OBD-II on vehicles since 1996, the diagnostic connectors are located in the same place on all vehicles. Scan tools have the ability to store, or "freeze" data during a road test, and then play back this data when the vehicle is returned to the shop.

There are many different scan tools available. Some are a combination of other diagnostic tools, such as a lab scope and graphing multimeter. These may have the following capabilities:

- Retrieve DTCs
- Monitor system operational data
- Reprogram the vehicle's electronic control modules
- Perform systems diagnostic tests
- Display appropriate service information, including electrical diagrams
- Display TSBs
- Troubleshooting instructions
- Easy tool updating through a personal computer (PC)

The vehicle's computer sets trouble codes when a voltage signal is entirely out of its normal range. The codes help technicians identify the cause of the problem when this is the case. If a signal is within its normal range but is still not correct, the vehicle's computer will not display a trouble code. However, a problem may still exist.

Some scan tools work directly with a PC through uncabled communication links, such as Bluetooth. Others use a Personal Digital Assistant (PDA). These are small hand-held units that allow you to read diagnostic trouble codes (DTCs), monitor the activity of sensors, and inspection/maintenance system test results to quickly determine what service the vehicle requires.

Most of these scan tools also have the ability to:

- Perform system and component tests
- Report test results of monitored systems
- Exchange files between a PC and PDA
- View and print files on a PC
- Print DTC/Freeze Frame
- Generate emissions reports
- IM/Mode 6 information
- Display related TSBs
- Display full diagnostic code descriptions
- Observe live sensor data
- Update the scan tool as a manufacturer's interfaces change
- Computer Memory Saver

Memory savers are an external power source used to maintain the memory circuits in electronic accessories and the engine, transmission, and body computers when the vehicle's battery is disconnected. The saver is plugged into the vehicle's cigar lighter outlet. It can be powered by a 9- or 12-volt battery. The air bag system must be disabled if the memory savers are used, they may have enough power to deploy the airbags while you are working on the vehicle.

Hybrid Tools

A hybrid vehicle is an automobile and as such is subject to many of the same problems as a conventional vehicle. Most systems in a hybrid vehicle are diagnosed in the same way as well. However, a hybrid vehicle has unique systems that require special procedures and test equipment. It is imperative to have good information before attempting to diagnose these vehicles. Also, make sure you follow all test procedures precisely as they are given.

An important diagnostic tool is a DMM. However, this is not the same DMM used on a conventional vehicle. The meter used on hybrids and other electric-powered vehicles should be classified as a category III meter. There are basically four categories for low-voltage electrical meters, each built for specific purposes and to meet certain standards. Low-voltage, in this case, means voltages less than 1000-volts.

The categories define how safe a meter is when measuring certain circuits. The standards for the various categories are defined by the American National Standards Institute (ANSI), the International Electrotechnical Commission (IEC), and the Canadian Standards Association (CSA). A CAT III meter is required for testing hybrid vehicles because of the high-voltages, three-phase current, and the potential for high transient voltages. Transient voltages are voltage surges or spikes that occur in AC circuits. To be safe, you should have a CAT III-1000 V meter. A meter's voltage rating reflects its ability to withstand transient voltages. Therefore, a CAT III 1000 V meter offers much more protection than a CAT III meter rated at 600 volts.

Another important tool is an insulation resistance tester. These can check for voltage leakage through the insulation of the high-voltage cables. Obviously no leakage is desired and any leakage can cause a safety hazard as well as damage to the vehicle. Minor leakage can also cause hybrid system-related driveability problems. This meter is not one commonly used by automotive technicians, but should be for anyone who might service a damaged hybrid vehicle, such as doing body repair. This should also be a CAT III meter and may be capable of checking resistance and voltage of circuits like a DMM.

To measure insulation resistance, system voltage is selected at the meter and the probes placed at their test position. The meter will display the voltage it detects. Normally, resistance readings are taken with the circuit de-energized unless you are checking the effectiveness of the cable or wire insulation. In this case, the meter is measuring the insulation's effectiveness and not its resistance.

The probes for the meters should have safety ridges or finger positioners. These help prevent physical contact between your fingertips and the meter's test leads.

Stethoscope

Some sounds can be easily heard without using a listening device, but others are impossible to hear unless they are amplified. A stethoscope is very helpful in locating the cause of a noise by amplifying the sound waves. It can also help you distinguish between normal and abnormal noise. The procedure for using a stethoscope is simple. Use the metal prod to trace the sound until it reaches its maximum intensity. Once the precise location has been discovered, the sound can be better evaluated. A sounding stick, which is nothing more than a long, hollow tube, works on the same principle, though a stethoscope gives much clearer results.

The best results, however, are obtained with an electronic listening device. With this tool you can tune into the noise. Doing this allows you to eliminate all other noises that might distract or mislead you.

Battery Hydrometer

On unsealed batteries, the specific gravity of the electrolyte can be measured to give a fairly good indication of the battery's state of charge. A hydrometer is used to perform this test. A battery hydrometer (Figure 6) consists of a glass tube or barrel, rubber bulb, rubber tube, and a glass float or hydrometer with a scale built into its upper stem. The glass tube encases the float and forms a reservoir for the test electrolyte. Squeezing the bulb pulls electrolyte into the reservoir.

When filled with test electrolyte, the sealed hydrometer float bobs in the electrolyte. The depth to which the glass float sinks in the test electrolyte indicates its relative weight compared to water. The reading is taken off the scale by sighting along the level of the electrolyte.

It is necessary to correct the reading by adding or subtracting 4 points (0.004) for each 10°F above or below the standard of 80°F. Battery testers are also available to test unsealed or sealed batteries that can check the condition of a battery electronically in just a few seconds. These testers can check the serviceability of a battery even if it is discharged.

Wire and Terminal Repair Tools

Many automotive electrical problems can be traced to faulty wiring. Loose or corroded terminals, frayed, broken, or oil soaked wires, and faulty insulation are the most common causes.

Wires and connectors are often repaired or replaced. Sometimes entire length of wires are replaced, other times only a section is. In either case, the wire must have the correct terminal or connector to work properly in the circuit. Wire cutters, stripping tools (Figure 7), terminal crimpers (Figure 8) and connector picks are the most commonly used tools for wire repair. Also, soldering equipment is used to provide the best electrical connection for a wire to another wire and for a wire to a connector.

Heat Gun

Often wire splices and connections are protected with heat shrink tubing. This tubing is slipped over the wire before repairs are made and then moved over the repair. A heat gun, which looks like a hair drier, is used to apply heat onto the tubing. The heat causes the tubing to shrink, forming a seal on the repair. The heat gun can also be used to simulate warm conditions on a component for testing purposes.

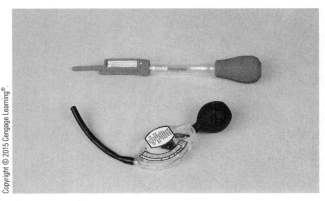

Figure 6 Battery hydrometers.

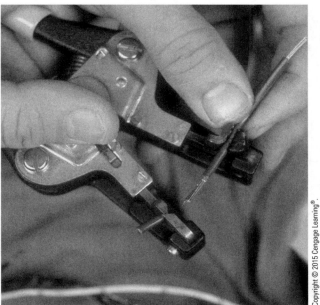

Figure 7 A wire-stripping tool.

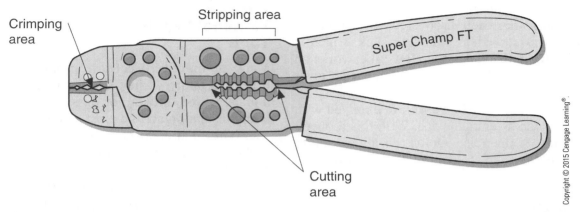

Figure 8 A wire terminal crimping tool.

Headlight Aimers

Headlights must be kept in adjustment to obtain maximum illumination. Sealed beams that are properly adjusted cover the correct range and afford the driver the proper nighttime view. Headlights can be adjusted using headlamp adjusting tools (Figure 9) or by shining the lights on a chart. Headlight aiming tools give the best results with the least amount of work. It should be noted that many late-model vehicles have levels built into the headlamp assemblies. These are used to correctly adjust the headlights.

Most headlight aimers use mirrors with split images, similar to split-image finders on some cameras, and spirit levels to determine exact adjustment. When using any headlight aiming equipment, follow the instructions provided by the equipment manufacturer.

Some late model vehicles headlamps are aligned by the use of a special screen that the headlamps are shined onto. The specifications for the screen are unique to the vehicle concerned.

Feeler Gauge

A feeler gauge is a thin strip of metal or plastic of known and closely controlled thickness. Several of these strips are often assembled together as a feeler gauge set that looks like a pocket-knife. A gauge of a desired thickness can be pivoted away from the others for convenient use. A feeler gauge set usually contains strips or leaves of 0.002- to 0.010-inch thickness (in steps of 0.001 inch) and leaves of 0.012- to 0.024-inch thickness (in steps of 0.002 inch). Metric feeler gauge sets are also available. These normally have leaves with a thickness of 0.02 to 1.00 millimeters.

A feeler gauge can be used by itself to measure clearances, such as critical compressor clutch clearances. It can also be used with a

Figure 9 Headlight aiming kit.

precision straightedge to measure the flatness of sealing surfaces in a compressor.

Straightedge

A straightedge is no more than a flat bar machined to be totally flat and straight and, to be effective, it must be flat and straight. Any surface that should be flat can be checked with a straightedge and feeler gauge set. The straightedge is placed across and at angles on the surface. At any low points on the surface, a feeler gauge can be placed between the straightedge and the surface. The size of the gauge that fills in the gap indicates the amount of warpage or distortion.

Dial Indicator

The dial indicator is calibrated in 0.001-inch (one-thousandth inch) increments. Metric dial indicators are also available. Both types are used to measure movement. Common uses of the dial indicator include checking compressor shaft runout, shaft endplay, and compressor clutch runout.

To use a dial indicator, position the indicator rod against the object to be measured. Then, push the indicator toward the work until the indicator needle travels far enough around the gauge face to permit movement to be read in either direction. Zero the indicator needle on the gauge. Move the object in the direction required, while observing the needle of the gauge. Always be sure the range of the dial indicator is sufficient to allow the amount of movement required by the measuring procedure. For example, never use a 1-inch indicator on a component that will move 2 inches.

Torque Wrench

Torque is the twisting force used to turn a fastener against the friction between the threads and between the head of the fastener and the surface of the component. The fact that practically every vehicle and engine manufacturer publishes a list of torque recommendations is ample proof of the importance of using proper amounts of torque when tightening nuts or bolts. The amount of torque applied to a fastener is measured with a torque-indicating or torque wrench.

A torque wrench is basically a ratchet or breaker bar with some means of displaying the amount of torque exerted on a bolt when pressure is applied to the handle. Torque wrenches are available with the various drive sizes. Sockets are inserted onto the drive and then placed over the bolt. As pressure is exerted on the bolt, the torque wrench indicates the amount of torque.

The common types of torque wrenches are available with inch-pound and foot-pound increments.

- A beam torque wrench is not highly accurate. It relies on a metal beam that points to the torque reading.
- A "click"-type torque wrench clicks when the desired torque is reached. The handle is twisted to set the desired torque reading.
- A dial torque wrench has a dial that indicates the torque exerted on the wrench. The wrench may have a light or buzzer that turns on when the desired torque is reached.
- A digital readout type displays the torque and is commonly used to measure turning effort, as well as for tightening bolts. Some designs of this type torque wrench have a light or buzzer that turns on when the desired torque is reached.

Gear and Bearing Pullers

Many tools are designed for a specific purpose. An example of a special tool is a gear and bearing puller. Many gears and bearings have a slight interference fit (press-fit) when they are installed on a shaft or in a housing. Something that has a press-fit has an interference fit. For example, if the inside diameter of a bore is 0.001 inch smaller than the outside diameter of a shaft, when the shaft is fitted into the bore it must be pressed in to overcome the 0.001 inch interference fit. This press-fit prevents the parts from moving on each other. The removal of these gears and bearings must be done carefully to prevent damage to the gears, bearings, or shafts. Prying or hammering can break or bind the parts. A puller with the proper jaws and adapters should be used to remove gears and bearings. Using the proper puller, the force required to remove a gear or bearing can be applied with a slight and steady motion.

There are pullers specifically designed for removing battery terminals from a battery

Figure 10 A battery terminal puller.

(Figure 10). These pullers should be used whenever the terminal doesn't slip off. Prying on the terminal to remove it can damage the battery.

Bushing and Seal Pullers and Drivers

Another commonly used group of special tools are the various designs of bushing and seal drivers and pullers. Pullers are either a threaded or slide hammer type tool. Always make sure you use the correct tool for the job; bushings and seals are easily damaged if the wrong tool or procedure is used.

Retaining Ring Pliers

A technician will run into many different styles and sizes of retaining rings and snap rings that hold things together or keep them in a fixed location. Using the correct tool to remove and install these rings is the only safe way to work with them. All technicians should have an assortment of snap ring and retaining ring pliers.

Door Panel Trim Tools

At times it is necessary to remove the interior door panels to diagnose or service electrical accessories. The door panels and some interior trim pieces are held in place with plastic clips. These clips can be difficult to remove without the proper tools. A variety of plastic and metal trim tools are available to remove the panels and trim. Using these will reduce the chance of harming the panels while removing them.

Service Information

Perhaps the most important tools you will use is service information. There is no way a technician can remember all of the procedures and specifications needed to correctly repair all vehicles. Therefore, a good technician relies on service information sources for this information. Good information plus knowledge allows a technician to fix a problem with the least bit of frustration and at the lowest cost to the customer.

To obtain the correct system specifications and other information, you must first identify the exact system you are working on. The best source for vehicle identification is the VIN. The code can be interpreted through information given in the service manual. The manual may also help you identify the system through identification of key components or other identification numbers and/or markings.

The primary source of repair and specification information for any car, van, or truck is the manufacturer. The manufacturer places service information on the Internet. The information is available for a subscription fee. Manufacturers' service information covers all repairs, adjustments, specifications, detailed diagnostic procedures, and the required special tools.

Since many technical changes occur on specific vehicles each year, manufacturers' service information need to be constantly updated The car manufacturer provides these bulletins to dealers and repair facilities on a regular basis, over the Internet.

Service information is also available from independent companies rather than the manufacturers. However, they pay for and get most of their information from the car makers. This information contains component information, diagnostic steps, repair procedures, and specifications for several car makes Information is usually condensed and is more general in nature than the manufacturer's manuals. The condensed format allows for more coverage in less space.

Many of the larger parts manufacturers have excellent guides on the various parts they manufacture or supply. They also provide updated service bulletins on their products. Other sources for up-to-date technical information are trade magazines and trade associations.

The same information that is available in service manuals is now commonly found electronically on compact disks, digital video disks (DVD), and online. Online data can be updated instantly and requires no space for physical storage. These systems are easy to use and the information is quickly accessed and displayed. The computer's keyword, mouse, and/or light pen are used to make selections from the screen's menu. Once the information is retrieved, a technician can read it off the screen or print it out and take it to the service bay.

JOB SHEETS
Required Supplemental Tasks (RST)

JOB SHEET 1

Shop Safety Survey

Name _____ Station _____ Date _____

NATEF Correlation

This Job Sheet addresses the following **RST** tasks:

Shop and Personal Safety:

1. Identify general shop safety rules and procedures.

8. Identify the location and use of eyewash stations.

10. Comply with the required use of safety glasses, ear protection, gloves, and shoes during lab/shop activities

11. Identify and wear appropriate clothing for lab/shop activities

12. Secure hair and jewelry for lab/shop activities.

Objective

As a professional technician, safety should be one of your first concerns. This job sheet will increase your awareness of shop safety rules and safety equipment. As you survey your shop area and answer the following questions, you will learn how to evaluate the safeness of your workplace.

Materials

Copy of the shop rules from the instructor

PROCEDURE

Your instructor will review your progress throughout this worksheet and should sign off on the sheet when you complete it.

1. Have your instructor provide you with a copy of the shop rules and procedures.

   Have you read and understood the shop rules?   ☐ Yes ☐ No

2. Before you begin to evaluate your work area, evaluate yourself. Are you dressed to work safely? ☐ Yes ☐ No

   If no, what is wrong? _____

3. Are your safety glasses OSHA approved? ☐ Yes ☐ No

   Do they have side protection shields? ☐ Yes ☐ No

4. Look around the shop and note any area that poses a potential safety hazard.

   _____

   _____

   All true hazards should be brought to the attention of the instructor immediately.

5. What is the air line pressure in the shop? _____ psi

   What should it be? _____ psi

6. Where is the first-aid kit(s) kept in the work area?

   _____

   _____

7. Ask the instructor to show the location of, and demonstrate the use of, the eyewash station. Where is it and when should it be used?

   _____

   _____

8. What is the shop's procedure for dealing with an accident?

   _____

   _____

   _____

9. Explain how to secure hair and jewelry while working in the shop.

   _____

10. List the phone numbers that should be called in case of an emergency.

   _____

Problems Encountered

_____

_____

_____

Instructor's Comments

_____

_____

_____

JOB SHEET 2

Working in a Safe Shop Environment

Name _____ Station _____ Date _____

NATEF Correlation

This Job Sheet addresses the following **RST** tasks:

Shop and Personal Safety:

2. Utilize safe procedures for handling of tools and equipment.

3. Identify and use proper placement of floor jacks and jack stands.

4. Identify and use proper procedures for safe lift operation.

5. Utilize proper ventilation procedures for working within the lab/shop area.

6. Identify marked safety areas.

9. Identify the location of the posted evacuation routes.

Objective

This job sheet will help you work safely in the shop. Two of the basic tools a technician uses are lifts and the floor jack.

This job sheet also covers the technicians' environment.

Materials

Vehicle for hoist and jack stand demonstration
Service information

Describe the vehicle being worked on:

Year _____ Make _____ Model _____

VIN _____ Engine type and size _____

PROCEDURE

1. Are there safety areas marked around grinders and other machinery? ☐ Yes ☐ No

2. Are the shop emergency escape routes clearly marked? ☐ Yes ☐ No

3. Have your instructor demonstrate the exhaust gas ventilation system in the shop. Explain the importance of the ventilation system.

   _____

   _____

4. What types of Lifts are used in the shop?

_____

_____

5. Find the location of the correct lifting points for the vehicle supplied by the instructor. On the rear of this sheet, draw a simple figure showing where these lift points are.

6. Ask your instructor to demonstrate the proper use of the lift.

   Summarize the proper use of the lift.

   _____

   _____

   _____

   _____

7. Demonstrate the proper use of jack stands with the help of your instructor.

   Summarize the proper use of jack stands.

   _____

   _____

   _____

Problems Encountered

_____

_____

_____

Instructor's Comments

_____

_____

_____

JOB SHEET 3

Fire Extinguisher Care and Use

Name _____ Station _____ Date _____

NATEF Correlation

This Job Sheet addresses the following **RST** task:

Shop and Personal Safety:

7. Identify the location and the types of fire extinguishers and other fire safety equipment; demonstrate knowledge of the procedures for using fire extinguishers and other fire safety equipment.

Objective

Upon completion of this job sheet, you will be able to demonstrate knowledge of the procedures for using fire extinguishers and other fire safety equipment, and identify the location of fire extinguishers in the shop.

NOTE: *Never fight a fire that is out of control or too large. Call the fire department immediately!*

1. Identify the location of the fire extinguishers in the shop.

   _____

   _____

2. Have the fire extinguishers been inspected recently? (Look for a dated tag.)

   _____

   _____

3. What types of fires are each of the shop's fire extinguishers rated to fight?

   _____

   _____

4. What types of fires should not be used with the shop's extinguishers?

   _____

   _____

5. One way to remember the operation of a fire extinguisher is to remember the term PASS. Describe the meaning of PASS on the following lines.

   a. P

   _____

   b. A

   _____

c. S

_____

d. S

_____

Problems Encountered

_____

_____

_____

Instructor's Comments

_____

_____

_____

JOB SHEET 4

Working Safely Around Air Bags

Name _____ Station _____ Date _____

NATEF Correlation

This Job Sheet addresses the following **RST** task:

Shop and Personal Safety:

13. Demonstrate awareness of the safety aspects of supplemental restraint systems (SRS), electronic brake control systems, and hybrid vehicle high-voltage circuits.

Objective

Upon completion of this job sheet, you should be able to work safely around and with air bag systems.

Tools and Materials

A vehicle(s) with air bag

Safety glasses, goggles

Service information appropriate to vehicle(s) used

Describe the vehicle being worked on:

Year _____ Make _____ Model _____

VIN _____ Engine type and size _____

PROCEDURE

1. Locate the information about the air bag system in the service information. How are the critical parts of the system identified in the vehicle?

   _____

   _____

2. List the main components of the air bag system and describe their location.

   _____

   _____

3. There are some very important guidelines to follow when working with and around air bag systems. Look through the service information to find the answers to the questions and fill in the blanks with the correct words.

   a. Wait at least _____ minutes after disconnecting the battery before beginning any service. The reserve _____ module is capable of storing enough energy to deploy the air bag for up to _____ minutes after battery voltage is lost.

b. Never carry an air bag module by its _____ or _____, and, when carrying it, always face the trim and air bag _____ from your body. When placing a module on a bench, always face the trim and air bag _____.

c. Deployed air bags may have a powdery residue on them. _____ is produced by the deployment reaction and is converted to _____ when it comes in contact with the moisture in the atmosphere. Although it is unlikely that harmful chemicals will still be on the bag, it is wise to wear _____ and _____ when handling a deployed air bag. Immediately wash your hands after handling a deployed air bag.

d. A live air bag must be _____ before it is disposed. A deployed air bag should be disposed of in a manner consistent with the _____ and manufacturer's procedures.

e. Never use a battery- or AC-powered _____, _____, or any other type of test equipment in the system unless the manufacturer specifically says to. Never probe with a _____ for voltage.

4. Explain how an air bag sensor should be handled before it is installed on the vehicle.

Problems Encountered

_____

_____

_____

Instructor's Response

_____

_____

_____

JOB SHEET 5

High-Voltage Hazards in Today's Vehicles

Name _____ Station _____ Date _____

NATEF Correlation

This Job Sheet addresses the following **RST** task:

Shop and Personal Safety:

14. Demonstrate awareness of the safety aspects of high-voltage circuits (such as high intensity discharge (HID) lamps, ignition systems, injection systems, etc.).

Objective

Upon completion of this job sheet, you will be able to describe some of the necessary precautions to take in performing work around high-voltage hazards such as HID headlamps and ignition systems.

HID Headlamp Precautions

Describe the vehicle being worked on:

Year _____ Make _____ Model _____

VIN _____ Engine type and size _____

Materials needed

Service information for HID headlamps

Describe general operating condition:

1. List three precautions a technician should observe when working with HID headlamp systems.

   _____

   _____

   _____

2. How many volts are necessary to initiate and maintain the arc inside the bulb of this HID system?

   _____

   _____

3. A technician must never probe with a test lamp between the HID ballast and the bulb. Explain why?

   _____

High-Voltage Ignition System Precautions

4. High-voltage ignition systems can cause serious injury, especially for those who have heart problems. Name at least three precautions to take when working around high-voltage ignition systems.

_____

_____

_____

_____

Problems Encountered

_____

_____

_____

Instructor's Comments

_____

_____

_____

JOB SHEET 6

Hybrid High-Voltage and Brake System Pressure Hazards

Name _____ Station _____ Date _____

NATEF Correlation

This Job Sheet addresses the following **RST** task:

Shop and Personal Safety:

13. Demonstrate awareness of the safety aspects of supplemental restraint systems (SRS), electronic brake control systems, and hybrid vehicle high-voltage circuits.

Objective

Upon completion of this job sheet, you will be able to describe some of the necessary precautions to take while performing work around high-voltage systems such as that used in hybrid vehicles, as well as the high-pressure systems of some braking control systems.

Tools and Materials

Appropriate service information

AUTHOR'S NOTE: *According to the vehicles' manufacturers, a technician should not work on hybrid vehicles without having the specific training, which is beyond the scope of this job sheet. This job sheet assumes that the student will be accessing information only, not actually working on live high-voltage vehicles.*

Protective Gear

Goggles or safety glasses with side shields

High-voltage gloves with properly inspected liners

Orange traffic cones to warn others in the shop of a high-voltage hazard

Describe the vehicle being worked on:

Year _____ Make _____ Model _____

VIN _____ Engine type and size _____

Describe general operating condition:

_____

PROCEDURE

High Pressure Braking Systems:

NOTE: *Many vehicles have a high pressure accumulator or a high pressure pump in their braking systems. Opening one of these systems can be hazardous due to the high pressures involved.*

1. Research the vehicle you have been assigned and describe the procedure that must be followed BEFORE opening the hydraulic braking system.

   _____

   _____

   _____

Hybrid Vehicles:

1. Special gloves with liners are to be inspected before each use when working on a hybrid vehicle.

   A. How should the integrity of the gloves be checked?

   _____

   _____

   B. How often should the gloves be tested (at a lab) for recertification?

   _____

   C. What precautions must be taken when storing the gloves?

   _____

   _____

   _____

   D. When must the gloves be worn?

   _____

   _____

   _____

2. What color are the high-voltage cables on hybrid vehicles?

   _____

3. What must be done BEFORE disconnecting the main voltage supply cable?

   _____

   _____

   _____

4. Describe the safety basis of the "one hand rule."

   _____

   _____

   _____

5. Explain the procedure to disable high voltage on the vehicle you selected.

   _____

   _____

   _____

6. Explain the procedure to test for high voltage to ensure that the main voltage is disconnected.

   _____

   _____

Problems Encountered

_____

_____

_____

Instructor's Comments

_____

_____

_____

JOB SHEET 7

Material Data Safety Sheet Usage

Name _____ Station _____ Date _____

NATEF Correlation

This Job Sheet addresses the following **RST** task:

Shop and Personal Safety:

12. Locate and demonstrate knowledge of material safety data sheets (MSDS).

Objective

On completion of this job sheet, the student will be able to locate the MSDS folder and describe the use of an MSDS sheet on the job site.

Materials

Selection of chemicals from the shop

MSDS sheets

1. Locate the MSDS folder in the shop. It should be in a prominent location. Did you have any problems finding the folder?

   _____

   _____

2. Pick a common chemical from your tool room such as brake cleaner. Locate the chemical in the MSDS folder. What chemical did you choose?

   _____

   _____

3. What is the flash point of the chemical? _____

4. Briefly describe why the flash point is important.

   _____

   _____

5. What is the first aid if the chemical is ingested?

   _____

   _____

6. Can this chemical be absorbed through the skin?

   _____

   _____

7. What are the signs of exposure to the chemical you selected?

   _____

   _____

8. What are the primary health hazards of the chemical?

   _____

   _____

9. What is the first aid procedure for exposure to this chemical?

   _____

   _____

10. What are the recommendations for protective clothing?

   _____

   _____

Problems Encountered

_____

_____

_____

Instructor's Comments

_____

_____

_____

JOB SHEET 8

Measuring Tools and Equipment Use

Name _____ Station _____ Date _____

NATEF Correlation

This Job Sheet addresses the following **RST** tasks:

Tools and Equipment:

1. Identify tools and their usage in automotive applications.

2. Identify standard and metric designations.

3. Demonstrate safe handling and use of appropriate tools.

4. Demonstrate proper cleaning, storage, and maintenance of tools and equipment.

5. Demonstrate proper use of precision measuring tools (i.e., micrometer, dial-indicator, dial caliper).

Objective

Upon completion of this job sheet, you will be able to make measurements using micrometers, dial indicators, pressure gauges, and other measuring tools. You will also be able to demonstrate the safe handling and use of appropriate tools, and demonstrate their proper use.

Tools and Materials

Items to measure selected by the instructor

Precision measuring tools: micrometer (digital or manual), dial caliper, vacuum or pressure gauge (selected by the instructor)

PROCEDURE

Have your instructor demonstrate the measuring tools that are available to you in your shop. The tools should include both standard and metric tools.

1. Describe the measuring tools you will be using.

   _____

   _____

2. Describe the items that you will be measuring.

   _____

   _____

3. Describe any special handling or safety procedures that are necessary when using the tools.

   _____

   _____

4. Describe any special cleaning or storage that these tools might require.

_____

_____

5. Describe the metric unit of measurement of the tools, such as millimeters, centimeters, kilopascals, etc.

_____

_____

6. Describe the standard unit of measure of the tools, such as inches, pounds per square inch, etc.

_____

_____

7. Measure the components, list them, and record your measurements in the following table.

Item measured	Measurement taken	Unit of measure

8. Clean and store the tools. Describe the process next.

Problems Encountered

_____

_____

_____

Instructor's Comments

_____

_____

_____

JOB SHEET 9

Preparing the Vehicle for Service and Customer

Name _____ Station _____ Date _____

NATEF Correlation

This Job Sheet addresses the following **RST** tasks:

Preparing Vehicle for Service:

1. Identify information needed and the service requested on a repair order.

2. Identify purpose and demonstrate proper use of fender covers, mats.

3. Demonstrate use of the three C's (concern, cause, and correction).

4. Review vehicle service history.

5. Complete work order to include customer information, vehicle identifying information, customer concern, related service history, cause, and correction.

Preparing Vehicle for Customer:

1. Ensure vehicle is prepared to return to customer per school/company policy (floor mats, steering wheel cover, etc.).

Objective

Upon completion of this job sheet, you will be able to prepare a service work order based on customer input, vehicle information, and service history. The student will also be able to describe the appropriate steps to take to protect the vehicle and delivering the vehicle to the customer after the repair.

Tools and Materials

An assigned vehicle or the vehicle of your choice

Service work order or computer-based shop management package

Parts and labor guide

Work Order Source: Describe the system used to complete the work order. If a paper repair order is being used, describe the source.

_____

_____

PROCEDURE

1. Prepare the shop management software for entering a new work order or obtain a blank paper work order. Describe the type of repair order you are going to use.

   _____

   _____

2. Enter customer information, including name, address, and phone numbers onto the work order.

   Task Completed ☐

3. Locate and record the vehicle's VIN. Where did you find the VIN?

   _____

4. Enter the necessary vehicle information, including year, make, model, engine type and size, transmission type, license number, and odometer reading.

   Task Completed ☐

5. Does the VIN verify that the information about the vehicle is correct?

   _____

6. Normally, you would interview the customer to identify his or her concerns. However, to complete this job sheet, assume the only concern is that the customer wishes to have the front brake pads replaced. Also, assume no additional work is required to do this. Add this service to the work order.

   Task Completed ☐

7. Prepare the vehicle for entering the service department. Add floor mats, seat covers, and steering wheel covers to the vehicle.

   Task Completed ☐

8. The history of service to the vehicle can often help diagnose problems, as well as indicate possible premature part failure. Gathering this information from the customer can provide some of the data needed. For this job sheet, assume the vehicle has not had a similar problem and was not recently involved in a collision. Service history is further obtained by searching files for previous service. Often this search is done by customer name, VIN, and license number. Check the files for any related service work.

   Task Completed ☐

9. Search for technical service bulletins on this vehicle that may relate to the customer's concern. Did you find any? _____ If so, record the reference numbers here.

   _____

   _____

10. Based on the customer's concern, service history, TSBs, and your knowledge, what is the likely cause of this concern?

    _____

    _____

11. Add this information to the work order.   Task Completed ☐

12. Prepare to make a repair cost estimate for the customer. Identify all parts that may need to be replaced to correct the concern. List these here.

_____

_____

13. Describe the task(s) that will be necessary to replace the part.

_____

_____

14. Using the parts and labor guide, locate the cost of the parts that will be replaced and enter the cost of each item onto the work order at the appropriate place for creating an estimate. If the valve or cam cover is leaking, what part will need to be replaced?

_____

15. Now, locate the flat rate time for work required to correct the concern. List each task with its flat rate time.

_____

_____

16. Multiply the time for each task by the shop's hourly rate and add the cost of each item to the work order at the appropriate place for creating an estimate. Ask your instructor which shop labor rate to use and record it here.

_____

17. Many shops have a standard amount they charge each customer for shop supplies and waste disposal. For this job sheet, use an amount of ten dollars for shop supplies.   Task Completed ☐

18. Add the total costs and insert the sum as the subtotal of the estimate.   Task Completed ☐

19. Taxes must be included in the estimate. What is the sales tax rate and does it apply to both parts and labor, or just one of these?

_____

_____

20. Enter the appropriate amount of taxes to the estimate, then add this to the subtotal. The end result is the estimate to give the customer.   Task Completed ☐

21. By law, how accurate must your estimate be?

_____

_____

22. Generally speaking, the work order is complete and is ready for the customer's signature. However, some businesses require additional information; make sure you add that information to the work order. On the work order, there is a legal statement that defines what the customer is agreeing to. Briefly describe the contents of that statement.

_____

23. Now that the vehicle is completed and it is ready to be returned to the customer, what are the appropriate steps to take to deliver the vehicle to the customer? What would you do to make the delivery special? What should not happen when the vehicle is delivered to the customer?

_____

_____

Problems Encountered

_____

_____

_____

Instructor's Comments

_____

_____

_____

JOB SHEETS

ELECTRICAL AND ELECTRONIC SYSTEMS JOB SHEET 10

Filling Out a Work Order

Name _____ Station _____ Date _____

NATEF Correlation

This Job Sheet addresses the following **MLR/AST/MAST** task:

A.1. Research applicable vehicle and service information, vehicle service history, service precautions, and technical service bulletins.

Objective

Upon completion of this job sheet, you will be able to research vehicle information, fluid types, vehicle service history, precautions, and technical bulletins

Tools and Materials

An assigned vehicle or the vehicle of your choice

Computer-based service information system

Service Information Source: Describe the system used to find service information.

_____

_____

_____

PROCEDURE

1. Enter customer information, including name, address, and phone numbers into the shop information/management system.   Task Completed ☐

2. Locate and record the vehicle's VIN. What is the VIN?

_____

3. Enter the necessary vehicle information, including year, make, model, engine type and size, transmission type, license number, and odometer reading.

Task Completed ☐

_____

4. Does the VIN verify that the information about the vehicle is correct?

_____

5. Normally, you would interview the customer to identify his or her concerns. However, to complete this job sheet, assume the only concern is that the right front headlamp does not work when the high beams are selected. This concern should be added to the work order.

Task Completed ☐

6. The history of service to the vehicle can often help diagnose problems as well as indicate possible premature part failure. Gathering this information from the customer can provide some of this information. For this job sheet, assume the vehicle has not had a similar problem and was not recently involved in a collision. Service history is further obtained by searching files for previous service. Often, this search is done by customer name, VIN, and license number. Check the files for any related service work.

Task Completed ☐

7. Search for technical service bulletins on this vehicle that may relate to the customer's concern. Were there any? If so, will they help with the repair?

_____

_____

8. Based on the customer's concern, service history, TSBs, and service information, what is the likely cause of this concern?

_____

_____

9. Describe the task(s) and precautions that will be necessary to replace the part.

_____

_____

10. Using the maintenance look-up tool in the service information system, describe the recommended fluid type for the following:

   A. Brake fluid

   _____

   B. Transmission fluid

   _____

C. Engine oil

_____

D. Power steering fluid

_____

Problems Encountered

_____

_____

_____

Instructor's Comments

_____

_____

_____

ELECTRICAL AND ELECTRONIC SYSTEMS JOB SHEET 11

Using Ohm's Law during Diagnosis

Name _____ Station _____ Date _____

NATEF Correlation

This Job Sheet addresses the following **MLR/AST/MAST** task:

A.2. Demonstrate knowledge of electrical/electronic series, parallel, and series-parallel circuits using principles of electricity (Ohm's law).

Objective

Upon completion of this job sheet, you will be able to use Ohm's law to help you logically diagnose electrical and electronic problems.

Tools and Materials

Paper and pen

PROCEDURE

> **NOTE:** *Studying Ohm's law is not merely an exhaustive exercise with numbers and formulas. Ohm's law describes how electrical circuits behave, and by knowing how a circuit should behave, it is easier to determine if something is wrong and what the basic problem is. This worksheet is not to be completed on a vehicle. Rather, it should be completed with you sitting down and thinking and applying your understanding of the principles of electricity to answer these questions.*

1. Briefly state what Ohm's law says. Do not enter the basic formula here.

   _____

   _____

2. Based on Ohm's law, what is the basic formula for finding amperage (current) when you know the voltage and resistance of the circuit?

   _____

   _____

3. What is the basic difference between a series circuit and a parallel circuit?

   _____

   _____

   _____

   _____

4. What is different about a series-parallel circuit? Give an example.

_____

_____

_____

_____

5. If the circuit resistance in a series circuit is increased, what happens to the circuit current?

_____

_____

6. If the circuit resistance in a parallel circuit is increased, what happens to the circuit current?

_____

_____

7. If the resistance of one of the resistors or loads in a series circuit increases, what happens to the circuit current?

_____

_____

8. If the resistance of one of the resistors or loads in a parallel circuit increases, what happens to the circuit current?

_____

_____

9. If there is a corroded ground in a series circuit, how are the circuit current and the voltage drops across the normal loads affected?

_____

_____

10. In a one-lamp light bulb circuit, if the bulb is burned out, what happens to the current in the circuit and how many volts would be dropped by the bulb?

_____

_____

11. What would happen to the operation of the circuit, as well as to the circuit current, if one leg (branch) of a parallel circuit is shorted to another leg of that circuit?

_____

_____

12. If one leg of a parallel circuit is shorted to ground, what happens to the other legs in that circuit? (Think deeply about this and talk about what happens before the circuit protection device opens!)

_____

_____

_____

_____

13. Fill-in-the-blanks: Based on Ohm's law and observation, we know that whenever there is an increase in circuit current, the resistance must have _____. This could only be caused by a(n) _____ because a(n) _____ would result in no current and a(n) _____ problem would result in a lower circuit current. We also know that when a light bulb burns dimmer than it should or a motor operates slower than it should, the cause must be _____ or lower source voltage. Further, we know if something doesn't work at all, the problem must be a(n) _____ or a(n) _____. If it is the latter, we will find an open circuit protection device or a burned wire.

Problems Encountered

_____

_____

_____

Instructor's Comments

_____

_____

_____

ELECTRICAL AND ELECTRONIC SYSTEMS JOB SHEET 12

Using a Digital Multimeter

Name _____ Station _____ Date _____

NATEF Correlation

This Job Sheet addresses the following **MLR** task:

A.4. Demonstrate proper use of a digital multimeter (DMM) when measuring source voltage, voltage drop (including grounds), current flow, and resistance.

This Job Sheet addresses the following **AST/MAST** task:

A.3. Demonstrate proper use of a digital multimeter (DMM) when measuring source voltage, voltage drop (including grounds), current flow, and resistance.

Objective

Upon completion of this job sheet, you will be able to use a digital multimeter for diagnosing electrical and electronic circuits.

Tools and Materials

Digital Multimeter (DMM)

Wiring diagrams

Service information

Protective Clothing

Goggles or safety glasses with side shields

Describe the vehicle being worked on:

Year _____ Make _____ Model _____

VIN _____ Engine type and size _____

PROCEDURE

Getting Familiar with Your DMM

1. List the manufacturer and model number of the DMM you are using for this worksheet.

_____

_____

2. Different DMMs are capable of measuring different things. What can this DMM do?

_____

_____

_____

3. In order to use this DMM to full capacity, do you need to insert a module or connect certain leads? If so, what are they?

_____

_____

_____

4. According to the literature that came with the DMM, what are the special features of this DMM?

_____

_____

_____

5. What is the impedance of the meter?

_____

_____

_____

6. How many decimal points are shown on the meter and what suffixes are used?

_____

_____

_____

7. What are the meter's low and high limits for measuring voltage, amperage, and resistance?

_____

_____

_____

Check Source Voltage on a Conventional 12-volt Battery.

> **NOTE:** *Do not measure the high-voltage battery in a hybrid/electric vehicle without the necessary factory training.*

1. Where is the battery of this vehicle located?

   _____

   _____

   _____

2. If the DMM has the autorange feature, set it to read DC voltage. If the DMM does not have autorange, set it to the proper voltage scale. Describe where you set the meter.

   _____

   _____

   _____

3. Connect the positive lead (the red lead) to the positive terminal of the battery.

   Task Completed ☐

4. Connect the negative lead (the black lead) of the meter to the negative terminal at the battery.

   Task Completed ☐

5. Record the voltage.

   _____

6. Check the wiring diagram and identify the power wire at the headlamp connector for low-beam operation. What color wire is it?

   _____

   _____

7. Move the positive test meter lead to that terminal and turn on the headlamps. Record the voltage. **(NEVER measure the voltage across a High Intensity Discharge (HID) lamp—the voltage is VERY high.)**

   _____

   _____

8. What does that voltage indicate? Was it the same as battery voltage?

   _____

   _____

Check Voltage Drop across Connectors

1. What connector will you be checking and what component is tied into that connector?

   _____

   _____

   _____

2. Turn on the ignition to supply power to the circuit.   Task Completed ☐

3. If the DMM has the autorange feature, set it to read DC voltage. If the
   DMM does not have autorange, set it to the proper voltage scale. Describe
   where you set the meter.

   _____

   _____

4. Connect the negative lead (the black lead) of the meter to a ground on the   Task Completed ☐
   vehicle.

5. Use the positive lead (the red lead) to check the voltage at one side of the   Task Completed ☐
   connector.

6. Turn on the circuit.   Task Completed ☐

7. Record the voltage.

   _____

8. Use the positive lead (the red lead) to check the voltage at the other side   Task Completed ☐
   of the connector.

9. Record the voltage.

   _____

10. Compare the reading taken in step 8 with the reading in step 10. What is
    indicated by the comparison?

    _____

    _____

11. If there is a difference, subtract the lower voltage from the higher voltage.
    Compare this to the specifications in the service information to determine
    if the voltage drop is within specifications. State your findings.

    _____

    _____

12. Move the negative meter lead to the power terminal on the side of the   Task Completed ☐
    connector that is closest to the component.

13. Connect the positive lead to the side of the connector that is closest to the Task Completed ☐
    battery.

14. Make sure the circuit is turned on and record the voltage.

    _____

15. That measurement is the voltage drop across the connector. Does this
    reading agree with the difference found in step 11? Explain your answer.

    _____

    _____

Using an Ammeter to Check Current Draw

1. When using a DMM to measure amperage, set the meter to the highest reading of amps provided by the meter. Why would you begin by doing this?

   _____

   _____

2. Describe the location of the windshield washer motor.

   _____

   _____

3. Using a wiring diagram, locate the wire that provides power to the windshield washer. Describe that wire.

   _____

   _____

4. If the ammeter has an inductive pickup (current probe), place the pickup clamp around the power wire for the washer motor.   Task Completed ☐

5. If the meter does not have an inductive pickup, the meter must be connected in series with the circuit. Locate a connector for the circuit and disconnect it. Connect fused jumper wires across the two halves of the connector to complete the circuit for all but the power circuit. How many terminals did you need to complete the circuit with jumper wires?

   _____

   _____

6. Connect the red meter lead to the terminal end farthest away from the motor, and then connect the opposite terminal.   Task Completed ☐

7. Activate the windshield washers and observe the reading on the meter. If there is no reading, select a lower amperage scale on the meter until you have a reading. What was the reading?

   _____

   _____

8. Compare your measurement to the specifications and give your conclusions about the test.

   _____

   _____

9. Troubleshoot that circuit for a short and repair the problem. After the problem has been corrected, test the battery drain again to verify the fix.   Task Completed ☐

Check Continuity with an Ohmmeter

NOTE: *An ohmmeter works by sending a small amount of current through the path to be measured. Because of this, all circuits and components being tested must be disconnected from power. An ohmmeter must never be connected to an energized circuit; doing so may damage the meter. The safest way to measure ohms is to disconnect the negative battery cable before taking resistance readings. NEVER USE AN OHMMETER TO CHECK RESISTANCE OF AN AIRBAG CIRCUIT.*

1. Locate the fuse panel or power distribution box.   Task Completed ☐

2. With no power to the fuses, check the resistance of each fuse. Summarize your findings:

   _____

   _____

   _____

3. Now connect the leads of the meter across the negative battery cable.

   Your reading is: _____ ohms

4. Disconnect the wires leading to the ignition coil primary connection.

   Now connect the leads of the meter across the terminals of the coil.

   Your reading is: _____ ohms

5. Reconnect the wires to the coil.

   Carefully remove one spark plug wire (if equipped) from the spark plug and the ignition coil or distributor cap.

   Now connect the leads of the meter across the wire.

   Your reading is: _____ ohms

   Carefully reinstall the spark plug wire

6. Refer to the service information and find out how to remove the bulb in the dome light.

   Remove it.

   Now connect the leads of the meter across the bulb.

   Your reading is: _____ ohms

7. Reinstall the bulb.

   Remove a rear brake light bulb.

   Now connect the leads of the meter across the bulb.

   Your reading is: _____ ohms

8. Reinstall the bulb

   Disconnect the wire connector to one of the headlights.

From the wiring diagram, identify which terminals are for low beam operation.

Now connect the leads of the meter across the low beam headlamp terminals.

Your reading is: _____ ohms

9. From the wiring diagram, identify which terminals are for high beam operation.

Now connect the leads of the meter across the high beam headlamp terminals.

Your reading is: _____ ohms

10. Check the service information to see if there are resistance specifications for the preceding items. If so, what were they? (Some of these items will not have specifications, but others will.)

_____

_____

_____

11. You measured the resistance across several different light bulbs. On each, you should have read a different amount of resistance. Based on your findings, which light bulb would be the brightest and which would be the dimmest? Explain why.

_____

_____

_____

Problems Encountered

_____

_____

_____

Instructor's Comments

_____

_____

_____

ELECTRICAL AND ELECTRONIC SYSTEMS JOB SHEET 13

Testing for Circuit Defects

Name _____ Station _____ Date _____

NATEF Correlation

This Job Sheet addresses the following **MLR** tasks:

A.5. Demonstrate knowledge of the causes and effects from shorts, grounds, opens, and resistance problems in electrical/electronic circuits.

A.7. Check operation of electrical circuits with fused jumper wires.

This Job Sheet addresses the following **AST/MAST** tasks:

A.4. Demonstrate knowledge of the causes and effects from shorts, grounds, opens, and resistance problems in electrical/electronic circuits

A.6. Check electrical circuits using fused jumper wires; determine necessary action.

Objective

Upon completion of this job sheet, you will be able to locate shorts, shorts-to-ground, opens, and high-resistance problems in electrical and electronic circuits.

Tools and Materials

Jumper wire   Test light

Cycling circuit breaker   Buzzer with fuse terminal adaptors

Digital multimeter (DMM)

Protective Clothing

Goggles or safety glasses with side shields

Describe the vehicle being worked on:

Year _____ Make _____ Model _____

VIN _____ Engine type and size _____

PROCEDURE

Testing for Opens

1. An open is usually indicated by an inoperative component or circuit. The easiest way to test a circuit is to start at the most accessible place and work from there. Is the load in the open circuit accessible?

   _____

2. Check for voltage at the input or positive side of the load. If the reading was 10.5 volts or higher, check the ground side of the load. If the voltage there is 1 volt or lower and the load did not work, the load is bad. If the

voltage at ground is greater than 1 volt, there is excessive resistance or an open in the ground circuit. If the voltage at the positive side of the load was less than 10.5 volts, proceed to the next step. But before you do, summarize your test results.

_____

_____

3. Move the positive lead of the voltmeter toward the battery, testing all connections along the way. If 10.5 volts or more are present at any connector, there is an open between that point and the point previously checked. Use a fused jumper wire to verify the location of the open. Summarize your test results.

_____

_____

4. If battery voltage was present at the ground of the load, there is an open in the ground circuit. If the voltage is over 0.5 volts on the ground side of the circuit, then there is excessive resistance in the ground circuit. Use a fused jumper wire to verify this. Summarize your test results.

_____

_____

Testing for Shorts

1. Use an ohmmeter to check for an internal short in a component. If the component is good, the meter will read the specified resistance or at least some resistance. If it is shorted, it will read lower than normal or zero resistance. Summarize your test results. *(Note: Never check for resistance at a module terminal unless directed to do so by service information.)*

_____

_____

_____

2. If the short is between circuits, check the wiring of the affected circuits for signs of burned insulation and melted conductors. Also, check common connectors that are shared by the two affected circuits. Summarize your test results.

_____

_____

_____

3. If a visual inspection does not identify the cause of a wire-to-wire short, remove one of the fuses for the affected circuits. Install a buzzer that has been fitted with the proper connectors across the fuse holder terminals. Activate the circuit the buzzer is connected to. Disconnect the loads that

are supposed to be activated by the switch. Then, disconnect the wire connectors in the circuit that connects the load to the switch. If the buzzer stops when a connector is disconnected, the short is in that circuit. Summarize your test results.

_____

_____

4. If the problem is a short to ground, the circuit's fuse or other protection device will be open. If the circuit is not protected, the wire, connector, or component will be burned or melted. As an aid to keep current flowing in the circuit so you can test it, connect a cycling circuit breaker across the fuse holder. The circuit breaker will continue to cycle open and closed, allowing you to test for voltage in the circuit. Is the fuse or circuit breaker open?

_____

5. Connect a test light in series with the cycling circuit breaker. While observing the test light, disconnect individual circuits and components one at a time until the light stays out. The short is in the circuit that was disconnected when the light went out. Summarize your test results.

_____

_____

_____

Testing for High Resistance

1. High-resistance problems are typically caused by corrosion on terminal ends, loose or poor connections, or frayed and damaged wires. Carefully inspect the affected circuit for these flaws. If the voltage drop is excessive, that part of the circuit contains the resistance. If the voltage drop was normal, the high resistance is in the switch or in the circuit feeding the switch. Summarize your test results.

_____

_____

_____

2. Whenever excessive resistance is suspected, both sides of the circuit should be checked. Begin by checking the voltage at the positive side of the load. This should be close to battery voltage unless the circuit contains a resistor located before the load. If the voltage is less than desired, check the voltage drop across the circuit from the switch to the load. If the voltage drop is excessive, that part of the circuit contains the resistance. If the voltage drop was normal, the high resistance is in the switch or in the circuit feeding the switch. Summarize your test results.

_____

_____

3. Check the voltage drop across the switch. If the voltage drop was excessive, the problem is the switch. If the voltage drop was normal, the high resistance is in the circuit feeding the switch. Summarize your test results.

_____

_____

4. If battery voltage was present at the load, the ground circuit for the load should be checked. Connect the red voltmeter lead to the ground side of the load and the black lead to the grounding point for the circuit. If the voltage drop was normal, the problem is the grounding point. If the voltage drop was excessive, move the black meter lead toward the red. Check the voltage drop at each step. Eventually you will read a high-voltage drop at one connector, and then a low drop at the next. The point of high resistance is between those two test points. If the voltage drop was excessive, that part of the circuit contains the resistance. If the voltage drop was normal, the high resistance is in the switch or in the circuit feeding the switch. Summarize your test results.

_____

_____

_____

Problems Encountered

_____

_____

_____

Instructor's Comments

_____

_____

_____

ELECTRICAL AND ELECTRONIC SYSTEMS JOB SHEET 14

Check Continuity in a Circuit with a Test Light

Name _____ Station _____ Date _____

NATEF Correlation

This Job Sheet addresses the following **MLR** task:

A.6. Check operation of electrical circuits with a test light.

This Job Sheet addresses the following **AST/MAST** task:

A.5. Check operation of electrical circuits with a test light.

Objective

Upon completion of this job sheet, you will be able to test continuity in circuits with a test light.

> **WARNING:** *Never use a regular test light on an electronic circuit; use only a high-impedance test light on these circuits. Never use a test light on airbag circuits.*

Tools and Materials
Test light
Wiring diagram
Service information

Protective Clothing
Goggles or safety glasses with side shields

Describe the vehicle being worked on:
Year _____ Make _____ Model _____

VIN _____ Engine type and size _____

PROCEDURE

1. Identify the circuit and component to be tested.

   _____

   _____

2. Describe the location of the accessible connectors for that component.

   _____

   _____

3. Is the circuit to be tested a power supply circuit or a ground circuit?

   _____

4. If the circuit to be tested is a power supply circuit, connect the clip end of the test light to a ground on the vehicle.

Task Completed ☐

5. Turn the key to the on position.

Task Completed ☐

6. With the light end, use the probe to check the wire in the circuit at each connector in the specific circuit. Describe what the light did at each wire. If the test light comes on, there is continuity between the power supply and the point where you are testing.

_____

_____

7. If the circuit to be tested is a ground circuit, connect the clip end of the test light to a power supply, such as the battery.

Task Completed ☐

8. With the light end, use the probe to check the wire in the circuit at each connector in the specific circuit. Describe what the light did at each wire. If the test light comes on, there is continuity between the ground and the point you are testing.

_____

_____

9. Describe your results.

_____

_____

Problems Encountered

_____

_____

_____

Instructor's Comments

_____

_____

_____

ELECTRICAL AND ELECTRONIC SYSTEMS JOB SHEET 15

Part Identification on a Wiring Diagram

Name _____ Station _____ Date _____

NATEF Correlation

This Job Sheet addresses the following **MLR** task:

> **A.3.** Use wiring diagrams to trace electrical/electronic circuits.

This Job Sheet addresses the following **AST/MAST** task:

> **A.7.** Use wiring diagrams during the diagnosis (troubleshooting) of electrical/electronic circuit problems.

Objective

Upon completion of this job sheet, you will be able to identify different parts on a wiring diagram to enable the diagnosis of electrical circuit problems.

Tools and Materials

A wiring diagram printed from an electronic service database (assigned by the instructor)

Describe the type of vehicle being worked on:

Year _____ Make _____ Model _____

Service manual used: _____

PROCEDURE

Look over the wiring diagram, and then answer the following questions:

1. Are all circuit grounds clearly marked in the wiring diagrams?

   _____

2. Are lines that cross each other always connected? How can you tell whether or not lines show a circuit connection?

   _____

3. Are switches shown in their normally open or normally closed position? Can you explain?

   _____

4. Do all wires have a color code listed by them?

   _____

5. Is the internal circuitry of all components shown in their schematic drawing?

   _____

List the location of the wiring diagram where the following electrical components are shown in the service information. Then, **draw the schematical symbol** used by this wiring diagram to represent the part.

Component	Location	Drawing
Windshield wiper motor	_____	_____
Dome (courtesy) light	_____	_____
A/C compressor clutch	_____	_____
Turn signal flasher unit	_____	_____
Fuse	_____	_____
Fuel gauge sending unit	_____	_____

Problems Encountered

_____

_____

_____

Instructor's Comments

_____

_____

_____

ELECTRICAL AND ELECTRONIC SYSTEMS JOB SHEET 16

Using an Ammeter to Check Parasitic Battery Drain

Name _____ Station _____ Date _____

NATEF Correlation

This Job Sheet addresses the following **MLR** task:

A.8. Measure key-off battery drain (parasitic draw).

This Job Sheet addresses the following **AST/MAST** task:

A.8. Diagnose the cause(s) of excessive key-off battery drain (parasitic draw); determine necessary action.

Objective

Upon completion of this job sheet, you will be able to check current flow in electrical and electronic circuits and components with an ammeter and measure and diagnose the causes of abnormal key-off battery drain.

Tools and Materials

Ammeter

Digital multimeter (DMM)

Protective Clothing

Goggles or safety glasses with side shields

Describe the vehicle being worked on:

Year _____ Make _____ Model _____

VIN _____ Engine type and size _____

PROCEDURE

1. If you are using a DMM to measure amperage, set the meter to the highest reading of amps provided by the meter. If using an ammeter, select the highest amperage scale available.   Task Completed ☐

2. If the ammeter has an inductive pickup for current measurements, place the pickup clamp around the wire in the circuit being tested. In this case, put it around the negative battery cable.   Task Completed ☐

3. If the meter does not have an inductive pickup, the meter must be connected in series with the circuit. Disconnect the negative cable of the battery. Connect the red meter lead to the cable end and the black meter lead to the battery.   Task Completed ☐

4. With everything turned off and the doors closed, what is the amperage reading shown on the meter? (Note: It can take a few minutes for all the electronics in the vehicle to shut down after the key has been turned off.)

_____

5. If there is no reading, select a lower amperage scale on the meter until you are able to take a reading. What was the reading?

_____

_____

6. Locate the allowable parasitic battery drain specification for the vehicle in the service manual. What is the maximum allowable amperage drain?

_____

_____

7. Compare your measurement to the specifications and give your conclusions about the test.

_____

_____

8. If the battery drain was excessive, visually check the trunk, glove box, and under hood lights to see if they are on. If any are on, remove the bulb and observe the ammeter. What did you observe?

_____

_____

9. If the drain is now within specifications, troubleshoot the removed light's circuit for a short.   Task Completed ☐

10. If no lights were found to be on, or the light was not the cause of excessive drain, go to the fuse panel or distribution center and remove one fuse at a time while watching the ammeter.   Task Completed ☐

11. When the battery drain decreases, the circuit protected by the fuse you removed last is the source of the problem. Describe what happened.

_____

_____

12. Troubleshoot that circuit for a short and repair the problem. After the problem has been corrected, test the battery drain again to verify the fix.   Task Completed ☐

Problems Encountered

_____

_____

_____

Instructor's Comments

_____

_____

_____

ELECTRICAL AND ELECTRONIC SYSTEMS JOB SHEET 17

Inspect and Check Fuses, Fusible Links, and Circuit Breakers

Name _____ Station _____ Date _____

NATEF Correlation

This Job Sheet addresses the following **MLR/AST/MAST** task:

A.9. Inspect and test fusible links, circuit breakers, and fuses; determine necessary action.

Objective

Upon completion of this job sheet, you will be able to inspect and test fusible links, circuit breakers, and fuses.

Tools and Materials

Digital Multimeter (DMM)

Test light

Service information

Protective Clothing

Goggles or safety glasses with side shields

Describe the vehicle being worked on:

Year _____ Make _____ Model _____

VIN _____ Engine type and size _____

PROCEDURE

1. Visually check fuses. Describe your observations.

   _____

   _____

   _____

2. Using a test light to test a fuse, ground the clip on one end of the test light. Touch the probe to each end of the fuse. If the light comes on at both ends of the fuse, it is good. If the light works only when touched to one end of the fuse, it is bad. Replace it. Describe your results.

   _____

   _____

   _____

3. Using a test light to test a fusible link, ground the clip on one end of the test light. Touch the probe to each end of the fusible link. If the light comes on at both ends of the fusible link, it is good. If the light only comes on at one end of the fusible link, it is bad. Replace it. Describe your results.

    _____

    _____

    _____

4. Using a DMM to test a fuse or fusible link can be done two different ways. If you are using the voltage scale, you will connect the black lead (the negative lead) to a ground. Touch the red lead (positive lead) to each end of the fuse or fusible link. Describe your results and state what is indicated by them.

    _____

    _____

5. To use an ohmmeter to test a fuse or circuit breaker, remove the fuse or circuit breaker. Touch the leads from the meter to each end of the fuse or circuit breaker. What was indicated on the meter and what does this indicate?

    _____

Problems Encountered

_____

_____

_____

Instructor's Comments

_____

_____

_____

ELECTRICAL AND ELECTRONIC SYSTEMS JOB SHEET 18

Testing Switches, Solenoids, Connectors, Relays, and Wires

Name _____ Station _____ Date _____

NATEF Correlation

This Job Sheet addresses the following **AST/MAST** task:

A.10. Inspect and test switches, connectors, relays, solenoid solid state devices, and wires of electrical/electronic circuits; determine necessary action.

Objective

Upon completion of this job sheet, you will be able to inspect and test switches, connectors, relays, solenoids, and wires of electrical and electronic circuits with a DMM and fused jumper wires.

Tools and Materials

Test light   Digital multimeter (DMM)

Fused jumper wire   Scan tool

Protective Clothing

Goggles or safety glasses with side shields

Describe the vehicle being worked on:

Year _____ Make _____ Model _____

VIN _____ Engine type and size _____

PROCEDURE

Protection Devices

1. Check a fuse with an ohmmeter. If the fuse is good, there will be continuity through it. To test a circuit protection device with a voltmeter, check for available voltage at both terminals of the unit. If the device is good, voltage will be present on both sides. A test light can be used in place of a voltmeter. Summarize the results of this check.

   _____

   _____

2. Measure the voltage drop across a fuse or other circuit protection device. If a fuse, a fuse link, or circuit breaker is in good condition, a voltage drop of zero will be measured. If 12 volts is read, the fuse is open. Any

reading between zero and 12 volts indicates some voltage drop. Make sure you check the fuse holder for resistance as well. Summarize the results of this check.

_____

_____

_____

Switches

1. To check a switch, disconnect the connector at the switch. With an ohm-meter, check for continuity between the terminals of the switch with the switch moved to the ON position and to the OFF position. With the switch in the OFF position, there should be no continuity between the terminals. With the switch on, there should be good continuity between the terminals. Summarize the results of this check.

_____

_____

_____

2. If the switch is activated mechanically and does not complete the circuit when it should, check the adjustment of the switch. If the adjustment is correct, replace the switch.

Task Completed ☐

3. Another way to check a switch is to simply bypass it with a fused jumper wire. If the component works when the switch is jumped, a bad switch is indicated. Summarize the results of this check.

_____

_____

_____

4. A voltage drop across switches should also be checked. Ideally, when the switch is closed, there should be no voltage drop or a very small one. Summarize the results of this check.

_____

_____

_____

5. If the switch has multiple throws and/or poles, refer to the wiring diagram to identify what terminals should be connected in each switch position. Check for continuity across the terminals. There should be low resistance across only those terminals that should be connected while the switch is in that position. Make sure to do this in all switch positions. Summarize the results of this check.

_____

_____

_____

Relays

1. Check the wiring diagram for the relay being tested to determine if the control is through an insulated or ground switch. If the relay is controlled on the ground side, continue with this procedure. If the relay is controlled on the positive side, describe the correct way to test it.

   _____

   _____

   _____

2. Connect the negative lead of the voltmeter to a good ground.   Task Completed ☐

3. Connect the positive lead to the output wire. If no voltage is present, continue testing. If there is voltage, disconnect the ground circuit of the relay. The voltmeter should now read 0 volts. If it does, the relay is good. If voltage is still present, the relay is faulty and should be replaced.   Task Completed ☐

4. Connect the positive voltmeter lead to the power input terminal. A reading near battery voltage should be obtained. If not, the power feed to the relay is faulty. If the correct voltage is present, continue testing.   Task Completed ☐

5. Connect the positive meter lead to the control terminal. A reading near battery voltage should be obtained. If not, check the circuit from the battery to the relay. If the correct voltage is there, continue testing.   Task Completed ☐

6. Connect the positive meter lead to the relay ground terminal. If more than 1 volt is present, the circuit has a poor ground. Summarize the results of this check.

   _____

   _____

   _____

Solenoids

1. Disconnect the connector to the solenoid.   Task Completed ☐

2. Measure resistance between the power terminal to the solenoid and body ground. What was your measurement and what does this indicate?

   _____

   _____

3. Connect a scan tool to the DLC.   Task Completed ☐

4. Determine how to activate the solenoid being tested. Briefly explain the procedure.

   _____

   _____

5. Activate the solenoid and listen. Did you hear a click? _____ If so, what does this indicate? If not, what does this indicate?

_____

_____

Stepped Resistors

1. Remove the resistor from its mounting.   Task Completed ☐

2. Make sure the ohmmeter is set to the correct scale for the anticipated   Task Completed ☐
   amount of resistance. Connect the ohmmeter leads to the two ends of the
   resistor.

3. Compare the results against specifications. Summarize the results of
   this check.

_____

_____

_____

Variable Resistors

1. Identify the input and output terminals and connect the ohmmeter across   Task Completed ☐
   them.

2. Rotate or move the variable control while observing the meter. The
   resistance value should remain within the limits specified for the resistor.
   If the resistance values do not match the specified amounts, or if there is
   a sudden change in resistance as the control is moved, the unit is faulty.
   Summarize the results of this check.

_____

_____

_____

Problems Encountered

_____

_____

_____

Instructor's Comments

_____

_____

_____

ELECTRICAL AND ELECTRONIC SYSTEMS JOB SHEET 19

Connector and Wire Repairs

Name _____ Station _____ Date _____

NATEF Correlation

This Job Sheet addresses the following **MLR/AST/MAST** task:

A.11. Replace electrical connectors and terminal ends.

This Job Sheet addresses the following **AST/MAST** task:

A.12. Repair wiring harness.

Objective

Upon completion of this job sheet, you will be able to repair wiring harnesses, terminal ends, and connectors.

Tools and Materials

Terminal repair kit   Heat shrink tubing
Crimping tool   Heat gun
Dielectric grease   Electrical tape

Protective Clothing

Goggles or safety glasses with side shields

Describe the vehicle being worked on:

Year _____ Make _____ Model _____

VIN _____ Engine type and size _____

PROCEDURE

1. Check all connectors and terminals inside the connector for corrosion, dirt, and looseness. Describe your findings.

   _____

   _____

2. Check the wiring harness for frayed, broken, or oil-soaked wires, and faulty insulation. Describe your findings.

   _____

   _____

CAUTION: *On some models, the SRS wires are in a separate harness. If the SRS harness is damaged, replace the harness, do not repair it. On other models, wire harnesses include yellow SRS wires. If any SRS wire is damaged, replace the entire harness; do not repair it.*

Removing a Terminal from a Connector

1. If a terminal needs to be replaced, it must first be removed from the connector. Carefully separate the two halves of the connector. Nearly all connectors have pushdown release type locks. Make sure these are not damaged when disconnecting the connectors. What did you find?

   _____

2. Refer to the service information for the correct way to release a terminal from the connector. Most connectors have a secondary lock that serves as a backup for the primary lock on the terminal. Make sure the procedure for releasing both locks is followed. Summarize that procedure.

   _____

   _____

   _____

Installing New Terminals

1. Carefully match the old terminal with a new one. Choose the correct replacement terminal based on the wire size range the terminal will accommodate.   Task Completed ☐

   WARNING: *Always follow the vehicle manufacturer's wiring and terminal repair procedure given in the service manual. On some components and circuits, manufacturers recommend component replacement rather than wiring repairs.*

2. Depending on the size of the wire you are repairing, use the proper size slot in the crimping tool.   Task Completed ☐

3. Strip the insulation off the end of the wire so the wire fits in the new terminal. If the wire has a wire seal, replace it with a new one. While stripping the wire, make sure not to cut any wire strands.   Task Completed ☐

4. Position the terminal in the crimping tool slot with the solid portion of the terminal toward the anvil.   Task Completed ☐

5. Insert the wire into the terminal. Sometimes it is necessary to twist the wire strands in order to insert the wire into the terminal.   Task Completed ☐

6. Squeeze the tool with both hands until the stops make contact.   Task Completed ☐

7. Inspect the quality of the wire crimp. If it has a poor crimp, cut it off and start over.   Task Completed ☐

8. Insert the terminal into the connector. Make sure the wire seals are pushed all the way into the connector. Then, pull gently on the wires to make sure the terminal is locked into place.   Task Completed ☐

9. Close or insert the secondary lock, if applicable, and reconnect the connector.   Task Completed ☐

Installing Pigtail Terminals

1. Pigtail terminals (short pieces of wire with a factory-crimped terminal) are used when the wire is too short or when access to the connector is too restricted to make a terminal repair. Remove the damaged or faulty terminal from the connector.

Task Completed ☐

2. Cut off the wire about an inch back from where it connects to the damaged or faulty terminal, then strip about half of the insulation off that piece.

Task Completed ☐

3. Select a pigtail terminal that matches the original wire at both ends (same kind of terminal and same diameter wire).

Task Completed ☐

4. Select the smallest splice connector that will fit onto the stripped end of the original wire.

Task Completed ☐

5. Insert the pigtail terminal into the connector; push it in until it locks in place.

Task Completed ☐

6. Lay the pigtail and the original wire side by side, and cut off both ends at once. When making more than one splice, do not cut each pigtail at the same location; the resulting "lump" of splice connectors would interfere with rewrapping the harness. Instead, cut the first pigtail close enough to the terminal to leave room to make each remaining cut about ¾ inch (20 mm) farther down on the next pigtail.

Task Completed ☐

7. Put the splice connector in the proper size slot in the crimper tool, slide it to one end (where the flare begins), and close the crimper handles far enough to hold it in place.

Task Completed ☐

8. Insert one of the bare wires into the splice connector end that is in the crimper. Push the wire all the way into the splice connector, and squeeze the crimper handles until the jaws touch.

Task Completed ☐

9. Crimp the other wire in the same way into the other end of the splice connector.

Task Completed ☐

10. After crimping, pull gently on the wires to make sure they are secure in the connector. Were you successful? If not, what should you do?

_____

_____

11. Separate the other wires in the harness from the repaired wire(s), and shield them with nonflammable material. What did you use to shield the other wires?

_____

_____

12. Apply heat from the heat gun at the middle of the splice connector, and move the gun toward the ends as the tube shrinks. Apply heat evenly by applying the heat around the splice connector. Shrinking is complete when a small amount of sealant appears at each end of the tube.

Task Completed ☐

Repairing Wires

> **NOTE:** *It may be necessary to bypass a length of wire that is not accessible. Cut the wire before it enters the inaccessible portion and at the other end where it leaves the area. Install a replacement wire and reroute it to the load. Be sure to protect the wire by using straps, hangers, or grommets as needed.*

> **WARNING:** *Use caution in rerouting wires when making repairs. Rerouting wires can result in induced voltages in nearby components. These stray voltages can interfere with the function of electronic circuits.*

1. Cut out the section of wire that needs to be replaced. Task Completed ☐

2. Cut a length of wire of the same size from a spool of wire. Cut the wire a few inches longer than the length cut out of the wiring harness. What size wire are you repairing? _____

3. Select the appropriate size and type of terminal connector. Be sure it is suitable for the unit's connecting post or prongs and it has enough current carrying capacity for the circuit. Also, make sure it is heavy enough to endure normal wire flexing and vibration. Task Completed ☐

4. Use the correct stripping opening on the crimping tool for the gauge of the wire and remove enough wire insulation to allow the wire to completely penetrate the connector. Strip both ends of the new wire and make sure the ends of the two wires (the circuit that is being repaired) remain in the wiring harness. Task Completed ☐

5. Place the replacement wire into the connector and crimp the connector. To get a proper crimp, place the open area of the connector facing toward the anvil. Make sure the wire is compressed under the crimp. Install a connector at both ends of the replacement wire. Task Completed ☐

6. If you are going to use heat shrink tubing, place the correct size of tubing over the ends of the replacement wire. Make sure the tubing is longer than the connector. Task Completed ☐

7. Insert the stripped ends of the wire in the harness into the connectors of the replacement wire and crimp them. Task Completed ☐

8. Use electrical tape or heat shrink tubing to seal the connection. If you are going to wrap the connection with tape, clean the area that will be taped with isopropyl alcohol. This will allow for good adhesion. Also, make sure the tape is tightly wrapped over the connections. Did you seal the connections with heat shrink tubing or tape?

Splicing Twisted or Shielded Wiring

1. Twisted and shielded wires are used in computer circuits to protect those circuits from electrical noise. To splice this type of wire, locate and cut out the damaged section of wire. Task Completed ☐

2. Do not cut into the Mylar tape or the drain wire. Remove about 1 inch (25 mm) of the outer jacket from the ends of the cable. Task Completed ☐

3. Unwrap the Mylar tape but do not remove it from the cable. Task Completed ☐

4. Untwist the wires and remove the insulation from the ends.   Task Completed ☐

5. Use a splice clip to connect the wires and solder the splice.   Task Completed ☐

6. Wrap the wires with the Mylar tape. Do not wrap the drain wire in the tape.   Task Completed ☐

7. If the drain wire is cut, splice the wire and solder the connection.   Task Completed ☐

8. Wrap the drain wire around the other wires and the Mylar tape.   Task Completed ☐

9. Use electrical tape or heat shrink tubing to insulate the cable.   Task Completed ☐

Problems Encountered

_____

_____

_____

Instructor's Comments

_____

_____

_____

ELECTRICAL AND ELECTRONIC SYSTEMS JOB SHEET 20

Soldering Two Copper Wires Together

Name _____ Station _____ Date _____

NATEF Correlation

This Job Sheet addresses the following **MLR** task:

A.10. Perform solder repair of electrical wiring.

This Job Sheet addresses the following **AST/MAST** task:

A.13. Perform solder repair of electrical wiring.

Objective

Upon completion of this job sheet, you will be able to make wire repairs using solder.

Tools and Materials

100-watt soldering iron	Splice clip
60/40 Rosin core solder	Heat shrink tube
Crimping tool	Heat gun

Protective Clothing

Goggles or safety glasses with side shields

PROCEDURE

1. Disconnect the fuse that powers the circuit being repaired. (**NOTE:** If a fuse does not protect the circuit, disconnect the ground lead of the battery.) What did you need to do?

   _____

   _____

2. Locate and cut out the damaged wire. Where is the wire and what circuit does it belong to?

   _____

   _____

3. Using the correct size stripper, remove about ½ inch of the insulation from both wires. What is the gauge of the wire?

   _____

4. Remove about ½ inch of the insulation from both ends of the replacement wire. The length of the replacement wire should be slightly longer than the length of the wire removed. What size wire are you inserting to make the repair?

_____

_____

5. Select the proper size splice clip to hold the splice.   Task Completed ☐

6. Place the correct size and length of heat shrink tube over the two ends of the wire.   Task Completed ☐

7. Overlap the two spliced ends and center the splice clip around the wires, making sure the wires extend beyond the splice clip in both directions.   Task Completed ☐

8. Crimp the splice clip firmly in place.   Task Completed ☐

9. Heat the splice clip with the soldering iron while applying solder to the opening of the clip. Do not apply solder to the iron. The iron should be 180 degrees away from the opening of the clip.   Task Completed ☐

10. After the solder cools, check the integrity of the joint. How did you do that?

_____

_____

11. Slide the heat shrink tube over the splice.   Task Completed ☐

12. Heat the tube with the hot air gun until it shrinks around the splice. Do not overheat the heat shrink tube. Describe any problems you faced.

_____

_____

_____

Problems Encountered

_____

_____

_____

Instructor's Comments

_____

_____

_____

ELECTRICAL AND ELECTRONIC SYSTEMS JOB SHEET 21

Observing and Analyzing Waveforms

Name _____ Station _____ Date _____

NATEF Correlation

This Job Sheet addresses the following **MAST** task:

A.14. Check electrical/electronic circuit waveforms; interpret readings and determine needed repairs.

Objective

Upon completion of this job sheet, you will be able to check electrical and electronic components by observing waveforms.

Tools and Materials

Graphing Multimeter (GMM) or Digital Storage Oscilloscope (DSO)

Instruction manual for the above tester

Wiring diagrams

Service information

Protective Clothing

Goggles or safety glasses with side shields

Describe the vehicle being worked on:

Year _____ Make _____ Model _____

VIN _____ Engine type and size _____

PROCEDURE

1. Describe the test equipment being used for this job sheet. Include manufacturer, model, and serial number.

   _____

   _____

2. Describe how the screen is divided: vertically and horizontally.

   _____

   _____

3. How many inputs can the tool handle at one time? How are the inputs labeled on the tool and what color are the leads?

   _____

   _____

4. Describe how the screen is adjusted to match the desired signal (position, range, time base, and triggering) and make it useable.

_____

_____

5. In general terms, describe what a waveform represents.

_____

_____

6. Locate the MAP sensor on the vehicle, and with a wiring diagram identify the wires at the MAP connector. Describe them here.

_____

7. Connect the scope to the MAP output and a good ground. When the engine is accelerated and returned to idle, the output voltage should smoothly increase and decrease. Describe your results and explain what is indicated by the waveform.

_____

_____

8. Connect the scope to the MAF sensor and measure the reference voltage and the integrity of the ground wire. Describe your results here.

_____

_____

_____

9. Locate the signal wire from the MAF sensor and connect the oscilloscope. Describe the waveform from the MAF sensor as you accelerate the engine from idle to around 2500 rpm.

_____

_____

10. Locate an oxygen sensor on the vehicle, and with a wiring diagram identify the wires at its connector. Describe them here.

_____

_____

11. Connect the lab scope to the sensor's signal wire and a good ground. Set the scope to display the trace at 200 millivolts per division and 500 milliseconds per division. It is usually best to run the engine at about 1500 rpm. The O_2 can be biased rich or lean, or not work at all, or work too slowly. Allow the engine and O_2 sensor to warm up. Watch the waveforms. If the sensor's voltage toggles between zero volts and 500 millivolts, it is toggling

within its normal range but it is not operating normally. It is biased low or lean. If the toggling only occurs at the higher limits of the voltage range, the sensor is biased rich. Describe what you observed.

_____

_____

Problems Encountered

_____

_____

_____

Instructor's Comments

_____

_____

_____

ELECTRICAL AND ELECTRONIC SYSTEMS JOB SHEET 22

Diagnosing LAN/CAN/BUS Wiring Systems

Name _____ Station _____ Date _____

NATEF Correlation

This Job Sheet addresses the following **MAST** task:

A.15. Repair CAN/BUS wiring harness.

Objective

Upon completion of this job sheet, you will be able to diagnose LAN/CAN/BUS wiring systems.

Tools and Materials

DMM

Scan tool

Vehicle equipped with a computer network

Protective Clothing

Goggles or safety glasses with side shields

Describe the vehicle being worked on:

Year _____ Make _____ Model _____

PROCEDURE

Look up and print out the wiring diagram for the network communications and answer the following questions.

1. How are the communication circuits (serial data) identified on the wiring schematic? Do they have a special designation, such as circuit 800?

   _____

2. How many modules are on the high-speed network? _____

3. How many modules are on the low-speed network? _____

4. Which terminals in the DLC are associated with the vehicle network? Identify the use of each of the following 16 terminals.

#1	#5	#9	#13
#2	#6	#10	#14
#3	#7	#11	#15
#4	#8	#12	#16

5. In the case of a CAN network, where are the termination resistors located? Are they part of a module, or are they a separate component installed in the wiring harness?

_____

_____

6. Briefly explain the function of the termination resistors on your vehicle.

_____

_____

7. Connect the scan tool, and using the procedure from your vehicle's service information and the scan tool operating guide, scan the vehicle for network codes and communication problems. List your results here:

_____

_____

_____

8. Using service information, see how many modules report to the scan tool. Does the number of high-speed and low-speed modules match the number found on the vehicle?

_____

9. If the vehicle network serial data lines are grounded, how would you begin to find the location of the grounded circuit?

_____

_____

_____

10. How could you tell if you found the location (or the approximate location) of the grounded circuit by watching the scan tool?

_____

_____

Problems Encountered

_____

_____

_____

Instructor's Comments

_____

_____

_____

ELECTRICAL AND ELECTRONIC SYSTEMS JOB SHEET 23

Inspection of a Battery for Condition, State of Charge, and Capacity

Name _____ Station _____ Date _____

NATEF Correlation

This Job Sheet addresses the following **MLR/AST/MAST** task:

B.1. Perform battery state-of-charge test (conductance); determine necessary action.

B.2. Confirm proper battery capacity for vehicle application; perform battery capacity test; determine necessary action.

Objective

Upon completion of this job sheet, you will be able to perform battery state-of-charge tests and battery capacity tests.

Tools and Materials

Snap-On MicroVat Tester or other conductance-type tester   Digital multimeter (DMM)

Starting/Charging System tester (VAT-40 or similar)   Service information

Protective Clothing

Goggles or safety glasses with side shields

Describe the vehicle being worked on:

Year _____ Make _____ Model _____

VIN _____ Engine type and size _____

PROCEDURE

1. Describe the general appearance of the battery.

   _____

   _____

2. Describe the manufacturer, model, and general ratings for the battery.

   _____

   _____

3. Using the appropriate service information, determine if this battery is the correct one for the vehicle. What did you find?

   _____

   _____

4. Describe the general appearance of the cables and terminals.

_____

_____

5. Check the tightness of the cables at both ends. Describe their condition.

_____

_____

6. Connect the positive lead of the DMM (set on DC volts) to the positive terminal of the battery.   Task Completed ☐

7. Put the negative of the meter on the battery case and move it all around the top and sides of the case. What readings did you get on the voltmeter?

_____

_____

8. What is indicated by those readings?

_____

_____

9. Measure the voltage of the battery. Your reading was _____ volts.

10. What do you know about the condition and state of charge of the battery based on the visual inspection and the preceding tests?

_____

_____

11. Locate the CCA rating of the battery. What is the rating? _____

12. If using a typical Starting/Charging System tester with an internal load, connect it to the battery. What type of tester will you be using?

_____

_____

13. Based on the CCA, how much load should be put on the battery during the capacity test? _____ amps

14. Conduct the battery load test. The battery's voltage decreased to _____ volts after _____ seconds.

15. If you are using a conductance-type tester, connect it and set it to conduct a battery test. Enter the CCA rating of the battery being tested. Start the test.   Task Completed ☐

16. Describe the results of either battery load (capacity) test. Include in the results your service recommendations and the reasons for them.

_____

_____

Problems Encountered

_____

_____

_____

Instructor's Comments

_____

_____

_____

ELECTRICAL AND ELECTRONIC SYSTEMS JOB SHEET 24

Maintaining Electronic Memory

Name _____ Station _____ Date _____

NATEF Correlation

This Job Sheet addresses the following **MLR/AST/MAST** task:

B.3. Maintain or restore electronic memory functions.

Objective

Upon completion of this job sheet, you will be able to maintain and restore electronic memory functions.

Tools and Materials

Memory saver

Protective Clothing

Goggles or safety glasses with side shields

Describe the vehicle being worked on:

Year _____ Make _____ Model _____

VIN _____ Engine type and size _____

PROCEDURE

1. Before disconnecting the battery on a vehicle, record all memory settings that the driver has control over, such as the radio and clock. Describe the accessories the vehicle has that have a memory function.

   _____

   _____

2. Insert the memory saver tool into the cigar lighter outlet or voltage receptacle.   Task Completed ☐

3. Disconnect the negative cable of the battery. Then, do whatever service work is required.   Task Completed ☐

4. Reconnect the negative battery cable.   Task Completed ☐

5. Remove the memory saver tool.   Task Completed ☐

6. Check the settings on the accessories noted in step 1.   Task Completed ☐

7. Refer to the service information and follow the procedures outlined for resetting the parameters of the computer. This normally consists of a test drive under prescribed operating modes. Summarize the requirements for the test drive.

_____

_____

_____

Problems Encountered

_____

_____

_____

Instructor's Comments

_____

_____

_____

ELECTRICAL AND ELECTRONIC SYSTEMS JOB SHEET 25

Remove, Clean, and Replace a Battery

Name _____ Station _____ Date _____

NATEF Correlation

This Job Sheet addresses the following **MLR** task:

B.4. Inspect and clean battery; fill battery cells; check battery cables, connectors, clamps, and hold-downs.

This Job Sheet addresses the following **AST/MAST** task:

B.4. Inspect, clean, fill, and replace battery, battery cables, connectors, clamps, and hold-downs.

Objective

Upon completion of this job sheet, you will be able to properly inspect, clean, fill, and replace a battery, as well as inspect and clean battery cables, connectors, clamps, and hold-downs.

Tools and Materials

Baking soda

Battery clamp puller

Battery cleaning wire brush

Battery strap or carrier

Wrench for battery terminal or cable-clamp pliers

Conventional wire brush and rags

Fender covers

Masking tape or bright felt marker

Petroleum jelly

Service information

Protective Clothing

Rubber gloves

Rubber apron

Face shield, goggles, or safety glasses with side shields

Describe the vehicle being worked on:

Year _____ Make _____ Model _____

VIN _____ Engine type and size _____

PROCEDURE

1. Place a fender cover around the work area.　　　　　　　　　　　　Task Completed ☐

2. Consult the appropriate service information for precautions about computer controls. What did you find?

_____

_____

3. Remove the negative (–) battery terminal. For connectors tightened with nuts and bolts, loosen the nut with a box wrench or cable-clamp pliers. Using ordinary pliers or an open-end wrench can cause problems. Always grip the cable while loosening the nut. This will prevent unnecessary pressure on the terminal post that could break it or loosen its mounting in the battery. If the connector does not lift easily off the terminal when loosened, use a clamp puller to free it. Prying with a screwdriver or bar strains the terminal post.   Task Completed ☐

4. Loosen the positive (+) battery cable and remove it from the battery with a battery terminal puller. If both battery cables are the same color, it is wise to mark the positive cable so that you will connect it to the correct terminal later. All that is needed to mark the cable is masking or a similar kind of tape. Although differences in terminal size are designed to prevent reversing the polarity of the battery, marking the positive cable is an extra safeguard.   Task Completed ☐

5. Loosen and remove the battery hold-down straps, cover, and heat shield.   Task Completed ☐

6. Lift the battery from the battery tray using a battery strap or carrier.   Task Completed ☐

7. Inspect and clean the battery tray with a solution of baking soda and water. What is the condition of the battery tray?

   _____

   _____

8. Inspect the battery and clean it with a solution of baking soda and water. What is the condition of the battery?

   _____

   _____

9. Clean the battery cable terminals with baking soda solution and a battery terminal brush. Use the external portion to clean the post and the internal portion for the terminal ends. What was the original condition of the cable ends and terminals?

   _____

   _____

10. Using a battery strap or carrier, install the replacement battery (new or recharged) in the battery tray.   Task Completed ☐

11. Install the battery cover or hold-down straps, and tighten their attaching nuts and bolts. Be certain the battery cannot move or bounce, but do not overtighten.   Task Completed ☐

12. If so equipped, reinstall the heat shield.   Task Completed ☐

13. Reinstall, beginning with the positive cable, both terminal connectors. Do not overtighten because this could damage the post or connectors. Coat the connectors with petroleum jelly.   Task Completed ☐

14. Test the installation by starting the engine.   Task Completed ☐

Problems Encountered

_____

_____

_____

Instructor's Comments

_____

_____

_____

ELECTRICAL AND ELECTRONIC SYSTEMS JOB SHEET 26

Charge a Battery

Name _____ Station _____ Date _____

NATEF Correlation

This Job Sheet addresses the following **MLR/AST/MAST** task:

B.5. Perform slow/fast battery charge according to manufacturer's recommendations.

Objective

Upon completion of this job sheet, you will be able to properly charge a battery.

Tools and Materials

Battery charger and cables and adapters
   for side terminal batteries

Battery clamp puller

Battery strap or carrier

Fender covers

Service information

Voltmeter

Protective Clothing

Gloves

Safety goggles, face shield, or safety glasses with side shields

PROCEDURE

CAUTION: *Special care must be given to charging maintenance-free batteries. If any electrolyte is lost during the charging process, the life of the battery is shortened. The fast-rate charge for this type of battery should be limited to 35 amperes for 20 minutes.*

1. Place fender covers around the work area. Remove battery from vehicle. It is also possible to charge a battery in the vehicle.

   Task Completed ☐

2. If performing an in-vehicle charge, consult the service information for any precautions associated with computer controls. What did you find?

   _____

   _____

3. Check that the charger is turned off. Connect the positive (+) cable from the charger to the positive (+) battery terminal. Connect the (–) cable from the charger to the negative (–) battery terminal. Be sure you have a good connection to prevent sparking.

   Task Completed ☐

4. If at no-load the battery reads below 12.2 volts, charge the battery according to the following rates:

Task Completed ☐

Battery Capacity (Reserve Minutes)	**Slow Charge**
80 minutes or less	10 hours at 5 amperes, or 5 hours at 10 amperes
Above 80–125 minutes	15 hours at 5 amperes, or 7½ hours at 10 amperes
Above 125–170 minutes	20 hours at 5 amperes, or 10 hours at 10 amperes
Above 170–250 minutes	30 hours at 5 amperes, or 15 hours at 10 amperes
Above 250 minutes	20 hours at 10 amperes

If the voltage of the battery at room temperature is 12.2 volts, charge the battery for half the time shown under "Slow Charge." If the voltage is 12.4 volts, charge the battery for one-fourth the "Slow Charge" time.

Turn the clock control on the charger to the desired charging time. Do not exceed the manufacturer's battery-charging limits, which generally appear on the battery. At what rate did you set the charger? Why?

_____

_____

5. Charge the battery to a voltage of at least 12.6 volts or until the green ball appears. Shaking or tipping the battery may be necessary to make the green ball appear. Never overcharge a battery.

Task Completed ☐

WARNING: *Never smoke around a charging battery. The hydrogen gas produced is highly explosive.*

6. When the battery is fully charged, turn off the power switch and disconnect the two battery charger cables from the battery. Return the charger to the proper area.

Task Completed ☐

Problems Encountered

_____

_____

_____

Instructor's Comments

_____

_____

_____

ELECTRICAL AND ELECTRONIC SYSTEMS JOB SHEET 27

Jump-Starting a Vehicle

Name _____ Station _____ Date _____

NATEF Correlation

This Job Sheet addresses the following **MLR/AST/MAST** task:

B.6. Jump-start vehicle using jumper cables and a booster battery or an auxiliary power supply.

Objective

Upon completion of this job sheet, you will be able to start a vehicle using jumper cables and a battery according to the manufacturer's recommendations.

Tools and Materials

Jumper cables

Protective Clothing

Goggles or safety glasses with side shields

Describe the vehicle being worked on:

Year _____ Make _____ Model _____

VIN _____ Engine type and size _____

PROCEDURE

1. To jump-start the vehicle, will you be using another vehicle or an auxiliary power source?

   _____

2. Where is the battery for the vehicle that needs to be jump-started?

   _____

3. If using another vehicle to jump-start, make sure the two vehicles are not touching each other.   Task Completed ☐

4. Apply the parking brake for each vehicle and put the transmissions in neutral or park.   Task Completed ☐

5. Turn off the ignition switch and the accessories on both vehicles.   Task Completed ☐

6. Attach one end of the positive jumper cable to the disabled battery's positive terminal.   Task Completed ☐

7. Connect the other end of the positive jumper cable to the booster battery's positive terminal.   Task Completed ☐

8. Attach one end of the negative jumper cable to the booster battery's negative terminal.

Task Completed ☐

9. Attach the other end of the negative jumper cable to an engine ground on the disabled vehicle.

Task Completed ☐

10. Attempt to start the disabled vehicle. If the disabled vehicle does not readily start, start the jumper vehicle and run it at fast idle.

Task Completed ☐

11. Once the disabled vehicle starts, disconnect the ground-connected negative jumper cable from the engine block.

Task Completed ☐

12. Disconnect the negative jumper cable from the booster battery.

Task Completed ☐

13. Disconnect the positive jumper cable from the booster battery, and then from the other battery.

Task Completed ☐

Problems Encountered

_____

_____

_____

Instructor's Comments

_____

_____

_____

ELECTRICAL AND ELECTRONIC SYSTEMS JOB SHEET 28

Disconnecting the High-Voltage Circuit on a Hybrid Vehicle

Name _____ Station _____ Date _____

NATEF Correlation

This Job Sheet addresses the following **MLR/AST/MAST** task:

B.7. Identify high-voltage circuits of an electric or hybrid electric vehicle and related safety precautions.

Objective

Upon completion of this job sheet, you will be able to locate and safely disconnect the high-voltage circuit on a hybrid vehicle.

Tools and Materials

Service information

Hybrid vehicle

Vinyl tape for insulation

Protective Clothing

Goggles or safety glasses with side shields

Insulated gloves that are dry and are not cracked, ruptured, torn, or damaged in any way

Describe the vehicle being worked on:

Year _____ Make _____ Model _____

VIN _____ Engine type and size _____

Describe the general condition of the vehicle:

_____

_____

PROCEDURE

> **WARNING:** *Unprotected contact with any electrically charged ("hot" or "live") high-voltage component could cause serious injury or death.*

1. Describe how this vehicle is identified as a hybrid.

_____

_____

2. After referring to the vehicle's owner's manual, briefly describe its operation.

_____

_____

_____

_____

3. How many volts are the batteries rated at?

_____

4. Where are the batteries located?

_____

_____

5. How are the high-voltage cables labeled and identified?

_____

_____

6. What is used to provide short circuit protection in the high-voltage battery pack?

_____

_____

7. What isolates the high-voltage system from the rest of the vehicle when the vehicle is shut off?

_____

_____

8. Using the vehicle's service and/or owner's manual as a guide, list at least five precautions that must be adhered to when servicing this hybrid vehicle.

_____

_____

_____

_____

_____

9. If the vehicle has been in an accident and some electrolyte from the batteries has leaked out, how should you clean up the spill?

_____

_____

_____

_____

10. Look under the hood and describe the components that are visible.

_____

_____

_____

_____

11. Without touching the high-voltage cables or components, describe the routing of the cables. Include what they appear to be connected from and to.

_____

_____

_____

_____

WARNING: _Never assume that a hybrid vehicle is shut off simply because it is silent. Make sure the ignition key is in your pocket and not in the ignition switch._

12. Describe the procedure for totally isolating the high-voltage system from the vehicle. This often involves the removal of a service plug. Be sure to include the location of this plug in your description of the procedure.

_____

_____

_____

_____

_____

_____

_____

_____

_____

13. Make two signs saying, "WORKING ON HIGH-VOLTAGE PARTS. DO NOT TOUCH!" Attach one to the steering wheel, and set the other one near the parts you are working on.   Task Completed ☐

14. After the service plug has been removed, how long should you wait before working around or on the high-voltage system?

_____

_____

_____

Problems Encountered

_____

_____

_____

Instructor's Comments

_____

_____

_____

ELECTRICAL AND ELECTRONIC SYSTEMS JOB SHEET 29

Memory Resets

Name _____ Station _____ Date _____

NATEF Correlation

This Job Sheet addresses the following **MLR/AST/MAST** task:

B.8. Identify electronic modules, security systems, radios, and other accessories that require reinitialization or code entry after reconnecting vehicle battery.

Objective

Upon completion of this job sheet, you will be able to identify what systems will be affected by a memory loss after the vehicle's battery has been disconnected. You will also be able to prepare for resetting the memory before disconnecting a battery.

Tools and Materials

Service manual

Protective Clothing

Goggles or safety glasses with side shields

Describe the vehicle being worked on:

Year _____ Make _____ Model _____

VIN _____ Engine type and size _____

Describe the accessories the vehicle is equipped with:

_____

_____

_____

_____

PROCEDURE

1. Depress the memory button for the driver's power seat. This will put the seat and possibly outside mirrors in the desired position. While working on the vehicle, do not change the position of the seat.

   Task Completed ☐

2. Turn on the radio. List all of the set AM stations with their button number:

   _____

   _____

3. List all of the FM radio stations with their button number (remember, many radios have multiple FM ranges, sometimes referred to as FM1, FM2, etc.). Make sure you record all set stations.

_____

_____

_____

_____

_____

_____

4. If the radio has an anti-theft feature, make sure you have the code. Where did you get the code? What is it?

_____

_____

_____

_____

5. Secure the code for the vehicle's anti-theft system. Check the owner's manual for the procedure for entering this code or for any special procedures that need to be followed when disconnecting the battery. Briefly describe the procedure.

_____

_____

6. On most vehicles, the battery can now be disconnected. Check the service information to see if anything else should be recorded prior to disconnecting the battery. Record your findings here:

_____

_____

7. Once the battery is reconnected, enter the vehicle's anti-theft code or follow the procedure for resetting the alarm system.

Task Completed ☐

8. Enter the radio anti-theft code. What is the procedure for doing this?

_____

_____

_____

9. Now enter AM and FM stations previously recorded. What is the procedure for doing this?

_____

_____

_____

10. Using an accurate source as a reference, set the time on the clock. What is the procedure for doing this?

   _____

   _____

11. Are there any other features of the radio or clock that need to be reset, such as refreshing a satellite radio?

   _____

   _____

12. If the power seat has not been moved since before the battery was disconnected, set the memory for the seat. What is the procedure for doing this?

   _____

   _____

13. Using the service information as a guide, determine if the memory of these other systems will need to be reset. Briefly describe the procedure for re-initializing the systems.

   A. Engine controls

   _____

   _____

   B. Idle learn

   _____

   _____

   C. Automatic transmission

   _____

   _____

   D. Suspension system

   _____

   _____

   E. Tire pressure monitoring system

   _____

   _____

   F. Power windows

   _____

   _____

G. Power hatch

_____

_____

H. Sliding roof

_____

_____

14. Are there any other precautions that must be followed before returning the vehicle to its owner?

_____

_____

Problems Encountered

_____

_____

_____

Instructor's Comments

_____

_____

_____

ELECTRICAL AND ELECTRONIC SYSTEMS JOB SHEET 30

Servicing the Auxiliary Battery in a Hybrid Vehicle

Name _____ Station _____ Date _____

NATEF Correlation

This Job Sheet addresses the following **MLR/AST/MAST** task:

B.9. Identify hybrid vehicle auxiliary (12v) battery service, repair, and test procedures.

Objective

Upon completion of this job sheet, you will be able to identify all special service procedures that relate to the servicing and testing of the auxiliary battery used in hybrid vehicles.

Tools and Materials

Service information

Hybrid vehicle

Protective Clothing

Goggles or safety glasses with side shields

Describe the vehicle being worked on:

Year _____ Make _____ Model _____

VIN _____ Engine type and size _____

PROCEDURE

1. According to the service information and/or owner's manual, how many batteries does this vehicle have? What are their voltage ratings?

   _____

   _____

2. Where are these batteries located?

   _____

   _____

3. What is the purpose of the auxiliary battery?

   _____

   _____

4. Are there any special service or test procedures required for this battery?
   If so, what are they?

   _____

   _____

Problems Encountered

_____

_____

_____

Instructor's Comments

_____

_____

_____

ELECTRICAL AND ELECTRONIC SYSTEMS JOB SHEET 31

Testing the Starter and Starter Circuit

Name _____ Station _____ Date _____

NATEF Correlation

This Job Sheet addresses the following **MLR/AST/MAST** tasks:

   C.1. Perform starter current draw test; determine necessary action.

   C.2. Perform starter circuit voltage drop tests; determine necessary action.

   C.3. Inspect and test starter relays and solenoids; determine necessary action.

Objective

Upon completion of this job sheet, you will be able to perform starter current draw tests and starter circuit voltage drop tests, as well as inspect and test starter relays and solenoids.

Tools and Materials

Fender covers	Digital multimeter (DMM)
Remote starter switch	Starting/Charging System tester (VAT-40 or similar)
Service information	Vise
Battery cables	Battery

Protective Clothing

Goggles or safety glasses with side shields

Describe the type of vehicle being worked on:

Year _____ Make _____ Model _____

VIN _____ Engine type and size _____

PROCEDURE

1. Draw a simple starter circuit for the vehicle. Use the wiring diagram as a guide. Make sure to include any relays and solenoids.

2. Place fender covers on the fenders of the vehicle.   Task Completed ☐

3. Disable the ignition or fuel injection to prevent the engine from starting.   Task Completed ☐

   Instructor's Check: _____

4. If available, connect a remote starter switch.   Task Completed ☐

5. Connect the voltmeter across the battery's negative cable. Crank the engine with the starter and observe the readings on the meter. Your reading was _____ volts.

6. What does this indicate?

   _____

   _____

7. Connect the voltmeter across the battery's positive cable (from battery to   Task Completed ☐
   starter motor).

   Crank the engine with the starter and observe the readings on the meter. Your reading was _____ volts.

8. This test measured the voltage drop across everything in the positive side of the circuit. What is included in this circuit?

   _____

   _____

   _____

9. What do the test results suggest? Where should the meter be connected to next to identify the cause of any excessive voltage drop?

   _____

   _____

   _____

10. Refer to the service manual for normal starter current draw.   Task Completed ☐

11. Expected starter current draw is _____ amps

    Voltage should not drop below _____ volts

12. Connect the Starting/Charging System tester cables to the vehicle.   Task Completed ☐

13. Zero the ammeter on the tester.   Task Completed ☐

14. Be prepared to observe the amperage when the engine begins to crank   Task Completed ☐
    and while it is cranking. Also note the voltage when you stop cranking the engine.

    The initial current draw was _____ amps

    After _____ seconds, the current draw was _____ amps and the voltage dropped to _____ volts.

15. What is indicated by the test results? Compare your measurements to the specifications.

_____

_____

_____

_____

_____

16. Reconnect the ignition or fuel injection system and start the engine.   Task Completed ☐

Free Speed (No-Load) Starter Motor Test

1. Remove the starter from the vehicle.   Task Completed ☐

2. Clamp the starter firmly in a bench vise.   Task Completed ☐

3. What are the no-load specifications for this starter motor?

_____

4. Connect the positive cable from the battery to the starter's battery connection.   Task Completed ☐

5. Put the inductive lead of the ammeter over the negative cable.   Task Completed ☐

6. Connect the negative cable from the battery to the frame of the starter.   Task Completed ☐

7. Is the motor running? Describe its sound.

_____

_____

8. Check the battery's voltage and the current draw and speed of the motor and compare them to specifications. Summarize the results.

_____

_____

_____

9. Based on the results of this test, what are your service recommendations?

_____

_____

_____

Problems Encountered

_____

_____

_____

Instructor's Comments

_____

_____

_____

ELECTRICAL AND ELECTRONIC SYSTEMS JOB SHEET 32

Remove, Inspect, and Replace a Starter Motor

Name _____ Station _____ Date _____

NATEF Correlation

This Job Sheet addresses the following **MLR/AST/MAST** task:

C.4. Remove and install starter in a vehicle.

Objective

Upon completion of this job sheet, you will be able to properly remove, disassemble, inspect, reassemble, and install a starter.

Tools and Materials

Appropriate screwdrivers	Lubricant
Appropriate wrenches	Rags
Cleaning solvent	Service information
Fender covers	Support brackets (if applicable)
Hoist or safety stands	Tags
Plastic head hammer	Punch
Snap-ring pliers	Scribe
Small press	Ball-peen hammer

Protective Clothing

Goggles or safety glasses with side shields

Describe the vehicle being worked on:

Year _____ Make _____ Model _____

VIN _____ Engine type and size _____

PROCEDURE

Starter Removal

1. Place fender covers on the vehicle. Disconnect the battery ground cable.   Task Completed ☐

   WARNING: *Never attempt to remove the starter without first disconnecting the battery.*

2. Raise and support the vehicle using a hoist or safety stands.   Task Completed ☐

3. If required to access the starter, remove the front wheels. Loosen or remove the exhaust pipes and other components that block access to the starter. What did you need to remove?

   _____

4. Disconnect and tag all wires from the starter motor and/or solenoid. Task Completed ☐

5. Remove the heat shields and support brackets. Remove all mounting bolts and shims that secure the starter to the engine, and lift out the starter. Task Completed ☐

6. Visually inspect the starter bushing and armature shaft bearing area. Inspect the pinion gear and overrunning clutch drive unit. Check the pinion housing for cracks. If a problem is found, replace the starter motor. Record your results on the Report Sheet for Starter Motor Inspection and Replacement. Task Completed ☐

7. Check the overrunning clutch pinion clearance by either forcing (where possible) the solenoid arm to the fully applied position or by energizing the starter solenoid. Task Completed ☐

8. When the solenoid has moved the pinion to the fully engaged position, push the pinion back to remove all slack. Measure the clearance between pinion and pinion stop. Record the results on the Report Sheet for Starter Motor Inspection and Replacement. Task Completed ☐

Name _____ Station _____ Date _____

REPORT SHEET FOR STARTER MOTOR INSPECTION AND REPLACEMENT		
	Serviceable	*Nonserviceable*
1. Visual inspection		
Bushing		
Armature shaft bearing		
Pinion gear		
Overrunning clutch		
Pinion housing		
Flywheel ring gear teeth		
	Clearance/Torque	
2. Actual pinion clearance		
3. Starter-motor-to-block-bolt torque		
Conclusions and Recommendations _____		
_____		
_____		

Starter Motor Disassembly and Reassembly

1. Clean the starter.   Task Completed ☐

2. Scribe reference marks at each end of the starter end housings and the   Task Completed ☐
   frame.

3. Disconnect the field coil connection at the solenoid. What terminal of the
   solenoid is this?

   _____

4. Remove the screws that attach the solenoid to the starter drive housing.   Task Completed ☐

5. Rotate the solenoid until the locking flange of the solenoid is free. Then,   Task Completed ☐
   remove the solenoid.

6. Remove the through-bolts from the end frame.   Task Completed ☐

7. Remove the end frame and frame of the starter.   Task Completed ☐

8. Remove the armature. What did you need to do in order to get the armature out?

   _____

   _____

9. Place a deep socket over the armature shaft until it contacts the retaining
   ring of the starter drive. What size socket did you use?

   _____

10. Tap the end of the socket with a plastic-faced hammer to drive the retainer   Task Completed ☐
    toward the armature. Move it only far enough to access the snap ring.

11. Remove the snap ring.   Task Completed ☐

12. Remove the retainer from the shaft and remove the clutch and spring   Task Completed ☐
    from the shaft. Press out the drive housing bushing and the end-frame
    bushing.

13. Inspect all parts and describe the condition of the following:

    Brushes: _____

    Armature: _____

    Commutator: _____

    Frame: _____

    Bearings: _____

    Starter drive: _____

    Field coils: _____

14. With an ohmmeter, check the armature for shorts, opens, and high resistance. How did you do that and what were your results?

_____

_____

_____

15. With an ohmmeter, check the field windings for shorts, opens, and high resistance. How did you do that and what were your results?

_____

_____

_____

16. Summarize your findings about the parts of the starter. What parts can be reused? What parts need to be replaced?

_____

_____

_____

17. Gather all of the required new parts, then press in the drive housing bushing and the end-frame bushing, using the appropriate drivers.

Task Completed ☐

18. Install the clutch and spring onto the armature shaft and install the retainer.

Task Completed ☐

19. Place the deep socket over the armature shaft until it contacts the retaining ring of the starter drive. Tap the end of the socket with a plastic-faced hammer to drive the retainer toward the armature. Move it only far enough to install a new snap ring.

Task Completed ☐

20. Install the armature into the frame. What did you need to do in order to get the armature in?

_____

_____

_____

21. Install the end frame over the armature shaft. Make sure the brushes are in contact with the commutator. Also make sure the end frame is aligned with the alignment marks you made during disassembly. What did you need to do in order to get the commutator past the brushes?

_____

_____

_____

22. Install and tighten the through-bolts from the end frame.   Task Completed ☐

23. Rotate the solenoid until it is locked in place.   Task Completed ☐

24. Install and tighten the screws that attach the solenoid to the starter drive   Task Completed ☐
    housing.

25. Reconnect the field coil connection at the solenoid.   Task Completed ☐

26. Bench test the starter. How did you do that and what were the results?

    _____

    _____

    _____

Installing a Starter

1. Lubricate the starter bushing and armature shaft with a few drops of   Task Completed ☐
   20-weight oil.

2. Attach the heat shield and support brackets to the new or serviced starter   Task Completed ☐
   motor. Some starters use a sealer in some locations to prevent entry of
   dust and water. When assembling parts, apply a nonhardening-type sealer.
   Where grommets are used, they must be in good condition and properly
   inserted.

3. Install the starter motor to the engine using all mounting bolts. Torque the
   bolts to specifications. What are the torque specifications?

   _____

4. Check for proper flywheel engagement. Many starter designs have no adjustment for pinion clearance. When it is wrong, correct it by installing
   new parts. Is there the proper pinion clearance, if not what is needed to
   correct it?

   _____

   _____

5. Reconnect all wires to the solenoid and/or motor terminals. Remount or   Task Completed ☐
   tighten any removed or loosened parts.

6. Lower the vehicle and reconnect the battery ground cable. Test the starter
   for proper operation. Does the starter work without abnormal noise?

   _____

   _____

7. Remove the fender covers. Clean work area and return vehicle to desig-   Task Completed ☐
   nated area.

Problems Encountered

_____

_____

_____

Instructor's Comments

_____

_____

_____

ELECTRICAL AND ELECTRONIC SYSTEMS JOB SHEET 33

Control Circuit Tests

Name _____ Station _____ Date _____

NATEF Correlation

This Job Sheet addresses the following **MLR/AST/MAST** task:

C.5. Inspect and test switches, connectors, and wires of starter control circuits; determine necessary action.

Objective

Upon completion of this job sheet, you will be able to inspect and test switches, connectors, and wires of starter control circuits.

Tools and Materials

Fender covers

Service information

Voltmeter

Protective Clothing

Goggles or safety glasses with side shields

Describe the vehicle being worked on:

Year _____ Make _____ Model _____

VIN _____ Engine type and size _____

PROCEDURE

1. Place fender cover on fender. Check vehicle wiring diagram, if possible, to identify all control circuit components. These normally include ignition switch, safety switch, starter solenoid winding, and/or a separate relay.

   Task Completed ☐

2. Disable the ignition or fuel injection system so the engine will not start when the starter is activated.

   Task Completed ☐

3. Connect the red lead of the voltmeter to the positive terminal of the battery and the black lead to the ignition switch terminal at the solenoid or starter relay.

   Task Completed ☐

4. Crank the engine with the starter and observe the voltage reading. Your reading was _____ volts.

5. What do you conclude from this test?

   _____

   _____

6. Normally, if you were diagnosing a system and you measured less than 0.1 volts during the last test, you would stop there and move on to testing another part of the starting system. For the purposes of this job sheet, proceed through the rest of the procedure.

7. Connect the voltmeter across the following circuits and record the voltage reading while the engine is being cranked.

   a. Battery (or fuse block) to ignition switch: _____ volts

   b. Input to output of the ignition switch: _____ volts

   c. Output of ignition switch to safety switch input: _____ volts

   d. Input of safety switch to output: _____ volts

   e. Output of safety switch to solenoid or relay: _____ volts

8. A reading of more than 0.1 volts across any one wire or switch usually indicates trouble. If a high reading is obtained across a safety switch used on automatic transmissions, check the adjustment of the switch according to the service information.   Task Completed ☐

9. Remove fender covers. Clean work area and return vehicle to designated area.   Task Completed ☐

Problems Encountered

_____

_____

_____

Instructor's Comments

_____

_____

_____

ELECTRICAL AND ELECTRONIC SYSTEMS JOB SHEET 34

Slow/No-Start Diagnosis

Name _____ Station _____ Date _____

NATEF Correlation

This Job Sheet addresses the following **AST/MAST** task:

> **C.6.** Differentiate between electrical and engine mechanical problems that cause a slow-crank or a no-crank condition.

Objective

Upon completion of this job sheet, you will be able to determine if the cause of a slow-start or a no-start condition is the engine or something in the starter or its electrical circuit.

Tools and Materials

Assorted Wrenches

Tachometer

Protective Clothing

Goggles or safety glasses with side shields

Describe the vehicle being worked on:

Year _____ Make _____ Model _____

VIN _____ Engine type and size _____

PROCEDURE

> **NOTE:** *Most of these questions can be answered whether or not the vehicle actually has a cranking problem. If it does not, don't answer the specific questions but be sure to answer the general questions.*

1. Prior to testing the starting system when there is a no-start or slow-start condition, certain checks should be made. The first of which is the state-of-charge of the battery. Measure the voltage of the battery and record your findings here.

   _____

   _____

2. Does this voltage rating indicate at least a three-quarter charged battery? If not, what should you do before continuing your diagnosis?

   _____

   _____

   _____

3. Connect a tachometer to the engine.   Task Completed ☐

4. Crank the engine and observe the tachometer. What did you observe?

   _____

5. Was the engine speed during cranking between 200–400 rpm? What does that indicate?

   _____

   _____

6. If the speed was below 200 or if the engine did not crank, what could be the problem?

   _____

   _____

   _____

   _____

7. If the engine speed was above 400 rpm, explain why.

   _____

   _____

   _____

   _____

8. If the engine did not crank, turn the crankshaft pulley in a clockwise direction with the appropriate wrench and/or socket. Describe your results.

   _____

   _____

9. What is indicated by the results of your tests? If the engine did not rotate or if it was very difficult to rotate it, what could be the cause?

   _____

   _____

10. If the engine had the problems you cited in answer #9, would that be enough to stop the starter motor from rotating the engine? Why?

   _____

   _____

Problems Encountered

   _____

   _____

   _____

Instructor's Comments

ELECTRICAL AND ELECTRONIC SYSTEMS JOB SHEET 35

Charging System Tests

Name _____ Station _____ Date _____

NATEF Correlation

This Job Sheet addresses the following **MLR** tasks:

D.1. Perform charging system output test; determine necessary action.

D.4. Perform charging circuit voltage drop tests; determine necessary action.

This Job Sheet addresses the following **AST/MAST** tasks:

D.1. Perform charging system output test; determine necessary action.

D.2. Diagnose charging system for the cause of undercharge, no-charge, and overcharge conditions.

D.5. Perform charging circuit voltage drop tests; determine necessary action.

Objective

Upon completion of this job sheet, you will be able to perform a charging system output test, diagnose a charging system for the cause of undercharge, no-charge, and overcharge conditions, as well as perform charging circuit voltage drop tests.

Tools and Materials

Service information and wiring diagram for the vehicle

Digital multimeter (DMM)

Starting/Charging System tester (VAT-40 or similar)

Protective Clothing

Goggles or safety glasses with side shields

Describe the vehicle being worked on:

Year _____ Make _____ Model _____

VIN _____ Engine type and size _____

PROCEDURE

1. Describe the general appearance of the AC generator and the wires that are attached to it.

   _____

   _____

   _____

   _____

2. Connect a voltmeter across the battery terminals. Task Completed ☐

3. Connect a tachometer if the vehicle doesn't have one. Task Completed ☐

4. With the engine off, measure and record the voltage at the battery. Describe your findings.

   _____

   _____

5. Check the alignment, condition, and tension of the generator's drive belt. Describe your findings.

   _____

   _____

6. Inspect the battery and the cables and terminals connected to it. Describe your findings.

   _____

   _____

7. Inspect all charging system wiring and connectors. Describe your findings.

   _____

   _____

8. Inspect the AC generator and regulator mountings for loose or missing bolts. Describe your findings.

   _____

   _____

9. Start the engine and bring its speed to about 1500 to 2000 rpm. Task Completed ☐

10. Measure and record the voltage at the battery. It should now be 13.5 to 14.5 volts. Describe your findings.

   _____

   _____

11. If the charging voltage was too high, describe what may be the cause.

   _____

   _____

12. If the charging voltage was too low, describe what may be the cause.

   _____

   _____

13. Explain how this test checks the operation of the voltage regulator.

_____

_____

_____

14. Identify the type and model of AC generator. What type is it?

_____

15. What are the output specifications for this AC generator?

    _____ amps and _____ volts

16. Connect the Starting/Charging System tester to the vehicle.

17. Start the engine and run it at the specified engine speed. What is that speed?

_____

18. Observe the output to the battery. The meter readings are:

    _____ amps and _____ volts

19. If readings are outside of the specifications, refer to the service information to find the proper way to full field the AC generator. Describe the method.

_____

_____

_____

_____

20. Full field the generator (if possible) and observe the output to the battery. (If it is not possible to full field the A/C generator, then refer to the service information for test procedures.) The meter readings are: _____ amps and _____ volts

21. Compare readings to specifications and give recommendations.

_____

_____

_____

_____

_____

22. Based on the preceding tests, what is indicated by the results?

_____

_____

_____

_____

_____

Problems Encountered

_____

_____

_____

Instructor's Comments

_____

_____

_____

ELECTRICAL AND ELECTRONIC SYSTEMS JOB SHEET 36

Inspect, Replace, and Adjust Drive Belts and Pulleys

Name _____ Station _____ Date _____

NATEF Correlation

This Job Sheet addresses the following **MLR** task:

D.2. Inspect, adjust, or replace generator (alternator) drive belts; check pulleys and tensioners for wear; check pulley and belt alignment.

This Job Sheet addresses the following **AST/MAST** task:

D.3. Inspect, adjust, or replace generator (alternator) drive belts; check pulleys and tensioners for wear; check pulley and belt alignment.

Objective

Upon completion of this job sheet, you will be able to inspect and adjust a generator (alternator) drive belt.

Tools and Materials

Two vehicles, one with a serpentine belt and the other with V-belts

Service information for the above vehicles

Belt tension gauge

Protective Clothing

Goggles or safety glasses with side shields

Describe the vehicles being worked on:

Year _____ Make _____ Model _____

VIN _____ Engine type and size _____

and

Year _____ Make _____ Model _____

VIN _____ Engine type and size _____

PROCEDURE

1. On the vehicle with a serpentine belt, carefully inspect the belt and describe the general condition of the belt.

_____

_____

_____

_____

2. With the proper belt tension gauge, check the tension of the belt. Note: most serpentine belt tensioners are marked to indicate the amount of wear on the serpentine belt. (Check Service Information) _____. You found _____.

3. Based on the preceding information, what are your recommendations?

   _____

   _____

4. Describe the procedure for adjusting the tension of the belt (if possible).

   _____

   _____

5. Referring to the service information, describe the procedure for removing the belt.

   _____

   _____

6. Draw the proper routing for installing a new drive belt.

7. Carefully inspect the belt(s) and describe the general condition of each one.

   _____

   _____

   _____

   _____

8. Based on the preceding information, what are your recommendations?

   _____

   _____

   _____

Problems Encountered

_____

_____

_____

Instructor's Comments

_____

_____

_____

ELECTRICAL AND ELECTRONIC SYSTEMS JOB SHEET 37

Remove, Inspect, and Replace an AC Generator

Name _____ Station _____ Date _____

NATEF Correlation

This Job Sheet addresses the following **MLR** task:

D.3. Remove, inspect, and install generator (alternator).

This Job Sheet addresses the following **AST/MAST** task:

D.4. Remove, inspect, and install generator (alternator).

Objective

Upon completion of this job sheet, you will be able to properly remove, inspect, and install a generator (alternator).

Tools and Material

400-grain polishing cloth	Press
Appropriate wrenches and screwdrivers	Pulley puller
Clean rags	Screwdrivers
Cleaning solvent	Scriber
Fender cover	Service information
High-temperature bearing grease	Small pry bar

Protective Clothing

Goggles or safety glasses with side shields

Describe the vehicle being worked on:

Year _____ Make _____ Model _____

VIN _____ Engine type and size _____

Describe the type of generator:

_____

PROCEDURE

Removing a Generator

1. Place fender covers on the fenders. Disconnect the negative (–) battery   Task Completed ☐
   cable first and then the positive (+) battery cable at the battery.

   WARNING: *Never attempt to remove the generator without isolating the battery first.*

2. Disconnect the wiring leads from the generator.   Task Completed ☐

3. Following the appropriate procedures, loosen the adjusting bolts and move the generator to provide sufficient slack to remove the generator drive belts.

Task Completed ☐

4. Remove the through-bolts that retain the generator.

Task Completed ☐

5. Remove the generator from the vehicle.

Task Completed ☐

Disassemble a Generator

1. Following the appropriate procedure, remove the brush assembly.

Task Completed ☐

2. Remove the pulley nut. Using a suitable puller, remove the pulley.

Task Completed ☐

   CAUTION: *Do not attempt to remove the pulley by prying it off.*

3. Remove the through-bolts. Mark the housing halves with a scriber. Pry the slip ring end frame and drive end frame apart, using a suitable small pry bar. Separate the units. Make sure the stator is pulled with the rear half.

Task Completed ☐

4. Inspect all parts. Using the service information as a guide, check the rotor, stator, and rectifier assembly with an ohmmeter. Record your results.

   _____

   _____

5. Based on these checks, describe the basic condition of the generator. Can it be reinstalled in the vehicle or should it be rebuilt or replaced? State your reasons for your recommendation.

   _____

   _____

Reassemble a Generator

1. Assemble in the reverse order of disassembly. Work carefully. Make sure all parts are correctly located. Align the scribe marks. If press fit is required, use a press.

Task Completed ☐

2. Torque the pulley retaining nut. Do not clamp the rotor to hold it while torquing the pulley nut. Doing so may deform the rotor. Clamp the pulley. What is the torque specification for the nut?

   _____

3. Reassemble the brush unit. Some generator brush assembly designs permit the use of a pin to hold the brushes in the holder during assembly. Some require hooking the brush leads over a section of the brush holder for installation. Another type uses simple brush holders that permit installation after the generator is assembled. If a brush holder pin is used, or if the leads are hooked, remove the pin or straighten them after assembly.

Task Completed ☐

4. Spin the rotor to check for free operation. Did it spin freely?

   _____

Installing a Generator

1. Install the generator onto its mounting bracket with bolts, washers, and nuts. Do not tighten.

   Task Completed ☐

2. Install the generator drive belts and tighten them to the proper belt tension.

   Specification _____

3. Tighten all generator retaining bolts to the specified torque.

   Specification _____

4. Install the generator terminal plug and battery leads to the unit.

   Task Completed ☐

5. Connect the negative battery cable.

   Task Completed ☐

6. Start the engine and check the generator for specified output and record:

   Specification _____

   Actual _____

   If the output is okay, remove the fender covers.

   Task Completed ☐

Problems Encountered

_____

_____

_____

Instructor's Comments

_____

_____

_____

ELECTRICAL AND ELECTRONIC SYSTEMS JOB SHEET 38

Diagnosing Light Circuits

Name _____ Station _____ Date _____

NATEF Correlation

This Job Sheet addresses the following **MLR** task:

E.1. Inspect interior and exterior lamps and socket including headlights and auxiliary lights (fog lights/driving lights); replace as needed.

This Job Sheet addresses the following **AST/MAST** tasks:

E.1. Diagnose (troubleshoot) the causes of brighter-than-normal, intermittent, dim, or no light operation; determine necessary action.

E.2. Inspect the interior and exterior lamps and socket, including headlights and auxiliary lights (fog lights/driving lights); replace as needed.

Objective

Upon completion of this job sheet, you will be able to diagnose the cause of a brighter than normal, intermittent, dim, or no light operation, as well as be able to inspect and diagnose turn signal or hazard light systems.

Tools and Materials

Wiring diagram for the vehicle

Digital multimeter (DMM)

Protective Clothing

Goggles or safety glasses with side shields

Describe the vehicle being worked on:

Year _____ Make _____ Model _____

VIN _____ Engine type and size _____

PROCEDURE

> **NOTE:** *This job sheet focuses on the turn signal and hazard light circuits. The logic and tech-niques used to determine the cause of various problems can be applied to all light circuits. Simply identify the parts of the circuit you need to test and match their function to the following components.*

1. Operate the turn signals for the right side of the vehicle and describe what happened.

_____

_____

2. Operate the turn signals for the left side of the vehicle and describe what happened.

_____

_____

3. Operate the hazard lights on the vehicle and describe what happened.

_____

_____

4. Check the interior lamps and see if they are all operating properly.

_____

_____

5. Does your assigned vehicle use a lighting module? If it does, which lamps does the module control?

_____

_____

6. If your vehicle uses a lighting module, do the turn signals and hazard flashers operate by using a flasher in the circuit?

_____

_____

7. If a lighting module is used on your assigned vehicle, how are the exterior lights diagnosed?

_____

_____

8. Match the problems with the following to diagnose the problem.

The bulbs burn brighter than normal.

- Check the other lights to see if they also burn brighter than normal. Check the voltage output of the charging system. Summarize what you found.

_____

_____

The bulbs work intermittently.

- If this problem only affects one bulb, check for a loose connection at that bulb socket or in that bulb's circuit. If only one side is affected, check for loose connections at the common points for those bulbs. If all bulbs are affected, check for a loose connection at the points that are common to all bulbs. Summarize what you found.

_____

_____

The bulbs are dimmer than normal.

- If this problem only affects one bulb, check for high resistance in the power and ground circuits for that bulb, as well as the bulb socket for damage or corrosion. If only one side is affected, check for high resistance at the common points for those bulbs. If all bulbs are affected, check for high resistance at the points that are common to all bulbs. Summarize what you found.

_____

_____

None of the lamps light.

- Check the fuse for that circuit. If the fuse is good, check for voltage at the common points in the circuit for all bulbs. Summarize what you found.

_____

_____

The hazards don't flash.

- Check the fuse or circuit breaker. If the fuse is good, suspect the hazard flasher unit and substitute a known good one for the suspected flasher unit. Check for an open in the circuit that is common to all hazard lights; this includes the switch. Summarize what you found.

_____

_____

The turn signals light but don't flash.

- Check the fuse or circuit breaker. If the fuse is good, suspect the turn signal flasher unit and substitute a known good one for the suspected flasher unit. Check for an open in the circuit that is common to all turn signal lights; this includes the switch and the circuit ground. Summarize what you found.

_____

_____

The turn signals flash at the wrong speed.

- Suspect the turn signal flasher unit (if applicable) and substitute a known good one for the suspected flasher unit. If the flasher was good, check the available voltage at the flasher unit. If it is too high, check the voltage output of the charging system. If it is too low, check for high resistance in the power circuit to the flasher. If the voltage is normal, check the operation of all bulbs. Summarize what you found.

_____

_____

The front turn signals don't light.

- Check for a loose connector or an open in the circuit that is only common to the front lights. Summarize what you found.

  _____

  _____

The rear turn signals don't light.

- Check for a loose connector or an open in the circuit that is only common to the rear lights. Summarize what you found.

  _____

  _____

One turn signal bulb doesn't light.

- Check the bulb. If the bulb is okay, check that lamp's circuit for an open or high resistance. Include a check of the ground circuit and also the bulb's socket. Summarize what you found.

  _____

  _____

Problems Encountered

_____

_____

_____

Instructor's Comments

_____

_____

_____

ELECTRICAL AND ELECTRONIC SYSTEMS JOB SHEET 39

Headlight Aiming

Name _____ Station _____ Date _____

NATEF Correlation

This Job Sheet addresses the following **MLR** task:

E.2. Aim headlights.

This Job Sheet addresses the following **AST/MAST** task:

E.3. Aim headlights.

Objective

Upon completion of this job sheet, you will be able to inspect, replace, and aim headlights and bulbs.

Tools and Materials

DMM

Wiring diagram for the vehicle

A vehicle with adjustable headlights

Portable headlight aiming kit or appropriate headlamp aiming screen

Protective Clothing

Goggles or safety glasses with side shields

Describe the vehicle being worked on:

Year _____ Make _____ Model _____

VIN _____ Engine type and size _____

PROCEDURE

1. Describe the type of headlights used on the vehicle.

   _____

   _____

2. Using the service information, identify what needs to be removed to replace a headlamp bulb. Briefly describe the procedure here.

   _____

   _____

3. According to service information, is it necessary to aim a headlamp assembly after headlamp bulb replacement. Describe your findings.

_____

_____

4. If the entire headlamp assembly housing needs to be replaced, will the headlamp system need to be aimed?

_____

5. What precautions must be adhered to when replacing the bulb?

_____

_____

Headlamp Aiming Procedure/Vehicles that Use Portable Aimers

1. Park the vehicle on a level floor.   Task Completed ☐

2. Install the calibrated aiming units to the headlights. Make sure the adapters fit the headlight aiming pads on the lens.   Task Completed ☐

3. Zero the horizontal adjustment dial. Are the split image target lines visible in the view port? _____

   If the lines cannot be seen, what should you do?

   _____

4. Turn the headlight horizontal adjusting screw until the split image target lines are aligned. Then, repeat this for the other headlight. List any problems you may have had doing this.

   _____

   _____

5. Turn the vertical adjustment dial on the aiming unit to zero.   Task Completed ☐

   Turn the vertical adjustment screw until the spirit level bubble is centered.

   Recheck your horizontal setting after adjusting the vertical.

6. List any problems you had in making the vertical adjustment.

   _____

   _____

Aiming Headlamps with a Screen

1. Describe the placement of the headlamp aiming screen for your assigned vehicle.

   _____

   _____

2. Describe the process to aim the headlamps with the screen.

_____

_____

3. Describe the horizontal adjustment of the headlamps.

_____

_____

_____

4. Describe the vertical adjustment process.

_____

_____

Problems Encountered

_____

_____

_____

Instructor's Comments

_____

_____

_____

ELECTRICAL AND ELECTRONIC SYSTEMS JOB SHEET 40

Working with High Intensity Discharge Headlights

Name _____ Station _____ Date _____

NATEF Correlation

This Job Sheet addresses the following **MLR** task:

E.3. Identify system voltage and safety precautions associated with high intensity discharge headlights.

This Job Sheet addresses the following **AST/MAST** task:

E.4. Identify system voltage and safety precautions associated with high intensity discharge headlights.

Objective

Upon completion of this job sheet, you will be able to safely work around and on high intensity discharge (HID) headlights by being aware of the special precautions that must be adhered to.

Tools and Materials

Service information

Protective Clothing

Goggles or safety glasses with side shields

Describe the vehicle being worked on:

Year _____ Make _____ Model _____

VIN _____ Engine type and size _____

PROCEDURE

1. How can you identify the high-voltage circuit in this HID system?

   _____

   _____

2. What should you do if a HID headlight has cracks, dents, chips, or other damage?

   _____

   _____

3. Does the HID system on this vehicle have a fail-safe mode? If so, what does it do?

   _____

   _____

4. How many volts may be present at the high-voltage socket for the head-lamp when the lamp is turned on?

_____

_____

5. What is inside the bulb and why is it important to be aware of the bulb's construction?

_____

_____

6. When servicing a HID headlight, what things should be done to avoid electrical shock?

_____

_____

_____

_____

_____

_____

_____

7. Why should you never turn on the headlights when there is no bulb in the headlamp?

_____

_____

8. Are there any other precautions given by the manufacturer that should be followed when servicing a HID? If so, what are they?

_____

_____

Problems Encountered

_____

_____

_____

Instructor's Comments

_____

_____

_____

ELECTRICAL AND ELECTRONIC SYSTEMS JOB SHEET 41

Checking Instrument Circuits

Name _____ Station _____ Date _____

NATEF Correlation

This Job Sheet addresses the following **AST/MAST** task:

F.1. Inspect and test gauges and gauge sending units for cause of abnormal gauge readings; determine necessary action.

Objective

Upon completion of this job sheet, you will be able to inspect and test gauges, gauge sending units, connectors, and wires for causes of abnormal gauge readings.

Tools and Materials

A vehicle with accessible sensors and switches

Component locator manual for the vehicle

A DSO

Digital multimeter (DMM)

Jumper wires

Service information

Protective Clothing

Goggles or safety glasses with side shields

Describe the vehicle being worked on:

Year _____ Make _____ Model _____

VIN _____ Engine type and size _____

PROCEDURE

1. Refer to the service manual for the proper procedure for running self-diagnostics on the instrument cluster. Summarize the procedure for doing this.

   _____

   _____

   _____

2. Summarize your findings from running self-diagnostics.

   _____

   _____

   _____

3. What are your service recommendations?

_____

_____

_____

4. If there were no trouble codes retrieved or all found faults were corrected, proceed to diagnose the system based on the following symptoms.

5. Some gauge readings are transmitted over serial data from other components. Looking at the wiring diagram from the service information, name two inputs that are sent from another module.

_____

_____

PROCEDURE

Some Gauge Readings Are Read as Direct Inputs to the Instrument Cluster. These Gauges Are Still Diagnosed Traditionally.

Gauge Reads Low Constantly

1. Using the wiring diagram from the service information, determine which gauge sending unit is wired directly to the instrument cluster. Describe your findings here.

_____

2. Disconnect the wire harness from the sending unit of the malfunctioning gauge. Where is this located?

_____

3. Connect a fused jumper wire between the gauge and the sending unit.   Task Completed ☐

4. Turn the ignition switch on and observe the gauge. Record what happened.

_____

_____

5. Explain why you connected the jumper wire. What were you looking for?

_____

_____

6. If the gauge reads too high with the jumper wire in place, visually inspect the ground for the sending unit and record your findings.

_____

_____

7. Check the ground with the DMM and record your findings.

_____

_____

8. If the ground was good but the gauge doesn't read correctly, what could be the cause?

_____

_____

9. If the gauge still reads low with the jumper wire, check the sending unit with an ohmmeter. Record your findings.

_____

_____

10. What are your service recommendations?

_____

_____

Gauge Reads High Constantly

1. Disconnect the wire harness from the sending unit of the malfunctioning gauge.   Task Completed ☐

2. Turn the ignition switch on and observe the gauge. Record what happened.

_____

_____

3. What would be the cause if the gauge now reads low?

_____

_____

4. If the gauge reads too high with the harness disconnected, visually inspect the wire from the sending unit to the gauge connector and record your findings.

_____

_____

5. Check for a short to ground in that circuit with the DMM and record your findings.

_____

_____

6. What are your service recommendations?

_____

_____

_____

Inaccurate Gauge Readings

1. Refer to the service information and identify the resistance values for the sending unit of the malfunctioning gauge. Record those specifications.

_____

_____

2. Remove the sending unit and check its resistance. Summarize your findings.

_____

_____

_____

3. Compare your measurements to the specifications. What are your service recommendations?

_____

_____

4. If the sending unit checked out fine, visually inspect all wires and connectors in the circuit and record your findings.

_____

_____

5. Check the wires and connectors with the DMM and record your findings.

_____

_____

6. What are your service recommendations?

_____

_____

_____

Bench Testing a Fuel Level Sending Unit

1. Locate the specifications for the fuel gauge sending unit in the service information. Record those specifications.

_____

_____

2. Select the ohmmeter function of the DMM.   Task Completed ☐

3. Connect the DMM's negative test lead to the ground terminal of the sending unit.   Task Completed ☐

4. Connect the meter's positive lead to the variable resistor terminal.   Task Completed ☐

5. Holding the sender unit in its normal position, place the float rod against the empty stop.   Task Completed ☐

6. Read the ohmmeter. What was the measurement?

_____

7. Compare that measurement to the specifications. What are your conclusions?

_____

_____

_____

8. Slowly move the float toward the full stop, while observing the ohmmeter. The resistance change should be smooth and consistent. Your measurement was: _____ What does this indicate?

_____

_____

9. Check the resistance value while holding the float against the full stop. What was the measurement?

_____

10. Compare that measurement to the specifications. What are your conclusions?

_____

_____

_____

11. Check the float to be sure it is not filled with fuel, and is not distorted or loose. Record your findings.

_____

_____

12. Summarize the results of this test.

_____

_____

_____

_____

Problems Encountered

_____

_____

_____

Instructor's Comments

_____

_____

_____

ELECTRICAL AND ELECTRONIC SYSTEMS JOB SHEET 42

Checking the Seat Belt Warning Circuitry

Name _____ Station _____ Date _____

NATEF Correlation

This Job Sheet addresses the following **AST/MAST** task:

F.2. Diagnose (troubleshoot) the causes of incorrect operation of warning devices and other driver information systems; determine necessary action.

Objective

Upon completion of this job sheet, you will be able to diagnose the cause of incorrect operation of warning devices and other driver information systems.

Tools and Materials

A vehicle with seat belts   Owner's manual for the above vehicle

Wiring diagram for the above vehicle   Digital multimeter (DMM)

Protective Clothing

Goggles or safety glasses with side shields

Describe the vehicle being worked on:

Year _____ Make _____ Model _____

VIN _____ Engine type and size _____

PROCEDURE

1 Describe the conditions during which the seat belt warning light and/or sound should be activated. Refer to the owner's manual.

_____

_____

2. Draw the seat belt warning circuit below. Identify the various controls of the circuit.

_____

_____

3. What other systems of the vehicle have warning devices? Base this on the wiring diagram and the owner's manual.

_____

_____

4. What conditions would cause these to illuminate or sound? Base this on the wiring diagram and the owner's manual.

_____

_____

_____

_____

5. Test the operation of each system by opening and closing each of the system controls. Record your observations and conclusions from these tests on the following lines.

_____

_____

_____

_____

_____

_____

Problems Encountered

_____

_____

_____

Instructor's Comments

_____

_____

_____

ELECTRICAL AND ELECTRONIC SYSTEMS JOB SHEET 43

Diagnosing Horn Problems

Name _____ Station _____ Date _____

NATEF Correlation

This Job Sheet addresses the following **AST/MAST** task:

G.1. Diagnose (troubleshoot) causes of incorrect horn operation; perform necessary action.

Objective

Upon completion of this job sheet, you will be able to diagnose incorrect horn operation.

Tools and Materials

Jumper wires   Digital multimeter (DMM)
Service information   Test light

Protective Clothing

Goggles or safety glasses with side shields

Describe the vehicle being worked on:

Year _____ Make _____ Model _____

VIN _____ Engine type and size _____

PROCEDURE

> **NOTE:** *Diagnosis of the horn circuit begins with the problem or customer's complaint. The procedures in this job sheet are divided by symptom. Choose the symptom that best describes the customer's complaint and follow that procedure. Although not all of the symptoms apply to one problem, it is wise to look through the procedures for all symptoms.*

Horn Doesn't Work

1. Verify the customer's complaint by depressing the horn button. Record your findings.

_____

_____

2. Rotate the steering wheel from stop to stop while depressing the horn button. Did the horn work? What are you checking for?

_____

_____

3. If the vehicle has an accessory remote key FOB, does the horn sound with the "panic" button? If the horn sounds with the panic button and not the horn pad, describe the next steps you would take in your diagnosis.

_____

_____

_____

_____

4. If the horn didn't work, check the fuse or fusible link. Describe its condition.

_____

_____

5. Did you test for power to the horn while an assistant pressed the horn button? _____ Describe your findings.

_____

_____

6. If there is no power to the horn terminal, use the schematic to check for a possible problem in the horn relay, power to the horn relay contacts, and ground to the appropriate contacts when the horn pad is depressed. Describe your findings here.

_____

_____

_____

7. If the test lamp lights at the horn terminal power connection, check the ground at the horn and record your findings.

_____

_____

Summarize your service recommendations.

_____

_____

_____

Poor Sound

1. Does the horn system have more than one horn? If yes, do both horns turn on when the horn button is depressed?

_____

_____

2. If one horn is not working, that will affect the overall quality of the horn sound. If this is the case, diagnose the cause of the inoperative horn and record your results.

_____

_____

3. Use a voltmeter to measure the voltage to the horn. What did you find?

_____

_____

4. If the voltage is less than what you would expect, check the voltage drops on the power side and ground side of the circuit. What did you find?

_____

_____

5. If no electrical problems were found, connect a jumper wire from the battery to the horn.   Task Completed ☐

6. Turn out the adjusting screw at the horn approximately one-quarter turn. Did the sound improve?

_____

_____

7. Adjust the screw as needed if the screw made a difference in sound quality. If the screw had no effect, replace the horn.   Task Completed ☐

Horn Sounds Continuously

1. Disconnect the horn relay from the circuit.   Task Completed ☐

2. With an ohmmeter, check for continuity from the battery terminal of the relay to the horn circuit terminal. What is indicated if there is continuity?

_____

_____

3. With an ohmmeter, check the continuity through the horn switch. What is indicated if there is continuity?

_____

_____

4. What are your service recommendations?

_____

_____

_____

Problems Encountered

_____

_____

_____

Instructor's Comments

_____

_____

_____

ELECTRICAL AND ELECTRONIC SYSTEMS JOB SHEET 44

Diagnosing a Windshield Wiper Circuit

Name _____ Station _____ Date _____

NATEF Correlation

This Job Sheet addresses the following **AST/MAST** task:

G.2. Diagnose (troubleshoot) causes of incorrect wiper operation; diagnose wiper speed control and park problems; perform necessary action.

Objective

Upon completion of this job sheet, you will be able to diagnose incorrect wiper operation and diagnose wiper speed control and park problems.

Tools and Materials

Wiring diagram for the vehicle

Service information for the vehicle

Digital multimeter (DMM)

Protective Clothing

Goggles or safety glasses with side shields

Describe the vehicle being worked on:

Year _____ Make _____ Model _____

VIN _____ Engine type and size _____

PROCEDURE

1. Describe the general operation of the windshield wipers. Check the operation in all speeds and modes.

   _____

   _____

   _____

2. Check the mechanical linkages for evidence of binding or breakage. Record your findings.

   _____

   _____

   _____

3. Draw the windshield wiper circuit below. Include its power source, controls, and ground.

4. Describe how the motor is controlled to operate at different speeds.

   _____

   _____

   _____

   _____

5. Connect the voltmeter across the ground circuit, and then energize the motor. What was your reading on the meter? What does this indicate?

   _____

   _____

   _____

   _____

6. Probe the power feed to the motor in the various switch positions (including the off or park position). Observe your voltmeter readings. What were they? What is indicated by these readings?

   _____

   _____

   _____

   _____

7. Describe the general operation of the windshield wiper system.

   _____

   _____

   _____

   _____

Problems Encountered

_____

_____

_____

Instructor's Comments

_____

_____

_____

ELECTRICAL AND ELECTRONIC SYSTEMS JOB SHEET 45

Diagnosing Windshield Washer Problems

Name _____ Station _____ Date _____

NATEF Correlation

This Job Sheet addresses the following **MLR** task:

F.5. Verify windshield wiper and washer operation; replace wiper blades.

Objective

Upon completion of this job sheet, you will be able to diagnose incorrect windshield washer operation.

Tools and Materials

Clean rag   Digital multimeter (DMM)

Catch basin   Service information

Protective Clothing

Goggles or safety glasses with side shields

Describe the vehicle being worked on:

Year _____ Make _____ Model _____

VIN _____ Engine type and size _____

PROCEDURE

1. Check the level of the fluid in the washer fluid container. Fill the container to the correct level, if necessary. What did you find?

   _____

   _____

2. Before beginning your diagnosis of the washer system, activate it and observe the sounds and actions of the system when you turn it on. Describe what you observed.

   _____

   _____

3. If the motor seems to run, check the lines for restrictions by removing the hose from the pump.   Task Completed ☐

4. Activate the system and describe what happened.

   _____

   _____

5. What are your service recommendations so far?

_____

_____

_____

6. If you determined there is a restriction somewhere in the fluid delivery or spray circuit, what can you do to identify the cause of the problem?

_____

_____

7. If the pump did not push out fluid or if the motor did not seem to run during the initial check, check the fuse and record its condition.

_____

_____

8. Activate the pump and look for signs that it is operating. What did you find?

_____

_____

9. If the pump works, check the fluid feed line for restrictions or damage. Record your results.

_____

_____

10. If the pump doesn't work, use a voltmeter to check for available voltage at the pump when it is activated. What did you find?

_____

_____

11. If there is power to the pump, check the pump's ground circuit. What did you find?

_____

_____

12. If the ground is good and there is power to the pump, the pump is bad and should be replaced.   Task Completed ☐

13. If there is no power to the pump, check for power to the switch. What did you find?

_____

_____

14. If there is power to the switch, check for power out of the switch. What did you find?

_____

_____

15. If there is power to the switch but no power out of the switch, replace the switch.   Task Completed ☐

16. If there is power in and out of the switch, check the power feed circuit to the pump. What did you find and what is the cause of the problem?

_____

_____

Problems Encountered

_____

_____

_____

Instructor's Comments

_____

_____

_____

ELECTRICAL AND ELECTRONIC SYSTEMS JOB SHEET 46

Diagnosing Motor-Driven Accessories

Name _____ Station _____ Date _____

NATEF Correlation

This Job Sheet addresses the following **AST/MAST** task:

H.1. Diagnose (troubleshoot) incorrect operation of motor-driven accessory circuits; determine necessary action.

Objective

Upon completion of this job sheet, you will be able to inspect and test A/C-heater blower motor circuits and components.

Tools and Materials

Wiring diagram for the vehicle
Digital multimeter (DMM)

Protective Clothing

Goggles or safety glasses with side shields

Describe the vehicle being worked on:

Year _____ Make _____ Model _____

VIN _____ Engine type and size _____

PROCEDURE

> **NOTE:** *Although this job sheet is focused on the blower motor, the procedures can be easily applied to all motor-driven accessories. Refer to the appropriate service manual and wiring diagram to identify the components of the circuit you wish to test. Then, apply the same sequence and logic given in this job sheet. Use the appropriate steps based on the type of blower motor control used on the vehicle.*

1. Refer to the service information or owner's manual to identify the proper operation of the blower motor. Is the blower motor operated automatically or is it manual?

   _____

   _____

2. Is the blower motor speed controlled by a resistor assembly and a switch, or is the speed adjusted by a control module?

   _____

   _____

3. Start the engine and turn on the blower. Move the blower control to all available positions and summarize what happened in each.

_____

_____

4. Match the type of circuit control and the problem with one of the following.

The blower works at some speeds but not all.

a. Check the voltage to the blower motor at the various switch positions.

b. If the voltage doesn't change when a new position is selected, check the circuit from the switch to the resistor block.

c. If there was zero voltage in a switch position, check for an open in the resistor block.

d. Give a summary of the test results.

_____

_____

The blower doesn't work and is controlled by a module. (Note: Use published diagnostics when possible.)

a. Check the fuse or circuit breaker. If it is open, check the circuit for a short to ground.

b. Check for power at the blower motor with a DMM. The power should increase with an increase in blower speed at the control. Summarize your results.

_____

c. If the voltage increases with an increase in blower speed command, check the ground at the blower motor. Summarize your results.

_____

d. If the blower motor did not operate, and power and ground are good, replace the blower motor and retest.

e. Give a summary of the test results.

_____

_____

The blower doesn't work and is controlled by a ground side switch.

a. Check the fuse or circuit breaker. If it is open, check the circuit for a short.

b. Connect a jumper wire from the motor's ground terminal to a known good ground. If the motor runs, the problem is in that circuit.

c. If the motor did not operate, check for voltage at the battery terminal of the motor. If no voltage is found, there is an open in the power feed circuit for the motor. If voltage was present, the motor is bad.

d. If the motor operated when the jumper wire was connected in step b, use an ohmmeter to check the ground connection of the switch.

e. If the ground is good, use a voltmeter to probe for voltage at any of the circuits from the resistor block to the switch. Replace the switch if there is power at that point.

f. If no voltage was available at the switch, check the circuit for an open. If there is not an open, suspect a faulty switch.

g. Give a summary of the test results.

_____

_____

Constantly running blower fan controlled by a module.

a. Using the vehicle schematics and circuit operation description, what is a possible cause of a constantly running blower fan? Summarize the possible causes next.

_____

_____

b. If the circuit is controlled by an insulated switch, check for a wire-to-wire short. Check other circuits of the vehicle to identify what circuit is involved in this problem. That circuit will also experience a lack of control, or when that circuit is turned off, the blower motor will turn off. The exact problem can be isolated by disconnecting portions of the circuit until the motor stops. The short is in that part of the circuit that disconnected last.

c. Give a summary of the test results.

_____

_____

_____

Problems Encountered

_____

_____

_____

Instructor's Comments

_____

_____

_____

ELECTRICAL AND ELECTRONIC SYSTEMS JOB SHEET 47

Diagnosing Power Lock Circuits

Name _____ Station _____ Date _____

NATEF Correlation

This Job Sheet addresses the following **AST/MAST** task:

H.2. Diagnose (troubleshoot) incorrect electric lock operation (including remote keyless entry); determine necessary action.

Objective

Upon completion of this job sheet, you will be able to diagnose incorrect electric lock operation.

Tools and Materials

Jumper wires
Digital multimeter (DMM)
Service information
Scan tool

Protective Clothing

Goggles or safety glasses with side shields

Describe the vehicle being worked on:

Year _____ Make _____ Model _____

VIN _____ Engine type and size _____

PROCEDURE

1. Briefly describe the operation of the power door locks on your assigned vehicle based on service information.

   _____

   _____

2. Check the operation of the key fob or remote door lock switch. What did you find?

   _____

   _____

3. Check the service information for instructions on recharging or replacing the battery in the remote unit. Describe those here.

   _____

   _____

If the remote unit's battery is good, but the remote does not lock or unlock any of the doors, and yet the locks do operate from the door switches, check the service information for possible causes. State your findings here.

_____

_____

4. If none of the door locks work from either the key FOB or the door lock switches, list the 3 possible causes here.

_____

_____

_____

5. If one door lock doesn't work, check to see if the manual door lock button is operational. If the manual door lock seems to be binding, check the door lock linkage. Describe what you found.

_____

_____

6. If no binding is detected, check the wiring connection to the door lock actuator.

Task Completed ☐

7. Check the service information to see which terminal should be connected to power and ground in both the lock and unlock positions. Using the DMM, describe your findings here.

_____

_____

8. If none of the door locks operate from one of the switches, but do from the other switch, check the continuity across the inoperative switch in all positions and record your findings. (Note: The scan tool can also be used to monitor the activity of the door lock switches on some vehicles.)

_____

_____

9. If the door switch is okay, or if all of the door locks don't work, remove the master switch.

Task Completed ☐

10. Check the continuity across the switch in all positions and record your findings.

_____

_____

11. Summarize your service recommendations for the door lock system.

_____

_____

_____

Problems Encountered

_____

_____

_____

Instructor's Comments

_____

_____

_____

ELECTRICAL AND ELECTRONIC SYSTEMS JOB SHEET 48

Speed or Cruise Control Diagnosis

Name _____ Station _____ Date _____

NATEF Correlation

This Job Sheet addresses the following **AST/MAST** task:

H.3. Diagnose (troubleshoot) incorrect operation of cruise control systems; determine necessary action.

Objective

Upon completion of this job sheet, you will be able to diagnose the incorrect operation of cruise control systems.

Tools and Materials

Scan tool

Service information

Protective Clothing

Goggles or safety glasses with side shields

Describe the vehicle being worked on:

Year _____ Make _____ Model _____

VIN _____ Engine type and size _____

WARNING: *Never attempt to drive a vehicle while operating a scan tool, or doing diagnostics. Always take someone to drive while you operate the scan tool. Only test drive according to your school policy, and it should always be away from heavy traffic. Always obey all traffic laws and operate the vehicle in a safe and responsible manner.*

PROCEDURE

1. Perform a visual inspection of the speed control system. Check the vehicle for trouble codes that might affect cruise control operation. Describe your results here.

   _____

2. Using the scan tool, check the cruise control switches operation. Do the switches change state on the scan tool as the switches are operated? Try the on/off, resume, and accelerate switches. Describe your results here.

   _____

   _____

Check the operation of the brake switch using the scan tool. Does it change state when the pedal is depressed?

_____

_____

CAUTION: *If at any time during the following steps the system should appear to go out of control and overspeed, be prepared to turn the system off at once.*

3. Accelerate and hold at 35 mph. Press and release the SET ACCEL button. *Hold foot pressure very lightly on the accelerator pedal until the system takes over. It should disengage immediately when the brake is applied.* Record your results on the Report Sheet for Cruise Control Road Test, found at end of this job sheet.   Task Completed ☐

4. Press the ON button, accelerate, and hold the speed at 35 mph. Press and hold the SET ACCEL button. Slowly remove your foot from the accelerator. The engine speed should gradually increase. Record your results on the Report Sheet for Cruise Control Road Test.   Task Completed ☐

5. When the speed reaches 50 mph, release the SET ACCEL button. The vehicle should maintain 50 mph. Record your results on the Report Sheet for Cruise Control Road Test.   Task Completed ☐

6. Press the COAST button and hold. The vehicle should start to decelerate. Slow the vehicle to 35 mph. Record your results on the Report Sheet for Cruise Control Simulated Road Test.   Task Completed ☐

7. Press and release the brake pedal. The system should shut off and the vehicle should decelerate. Record your results on the Report Sheet for Cruise Control Simulated Road Test.   Task Completed ☐

8. Accelerate the engine and set the speed at 50 mph. Brake to 35 mph and maintain 35 mph with the accelerator. Depress and release the RESUME button. The speed should return to 50 mph. Record your results on the Report Sheet for Speed Control Simulated Road Test.   Task Completed ☐

Problems Encountered

_____

_____

_____

Instructor's Comments

_____

_____

_____

Name _____ Station _____ Date _____

REPORT SHEET FOR SPEED CONTROL ROAD TEST		
	OK	*Service Required*
1. **Visual inspection**		
2. **SET ACCEL**	Yes	No
Hold at 35 mph		
Idle when OFF		
Increase from 35 mph		
Maintain 50 mph		
3. **Set COAST**		
Vehicle decelerates		
Maintain 35 mph		
4. **Depress brake pedal**		
Vehicle decelerates		
5. **Set RESUME**		
Return to 50 mph		
Conclusions and Recommendations _____		

ELECTRICAL AND ELECTRONIC SYSTEMS JOB SHEET 49

Safely Diagnosing Air Bag (SRS) Systems

Name _____ Station _____ Date _____

NATEF Correlation

This Job Sheet addresses the following **AST/MAST** task:

H.4. Diagnose (troubleshoot) supplemental restraint system (SRS) problems; determine necessary action.

Objective

Upon completion of this job sheet, you will be able to safely diagnose supplemental restraint system (SRS) concerns.

Tools and Materials

A vehicle with air bags
Service information for the above vehicle
Component locator for the above vehicle
Safety glasses
Digital multimeter (DMM)

Protective Clothing

Goggles or safety glasses with side shields

Describe the vehicle being worked on:

Year _____ Make _____ Model _____

VIN _____ Engine type and size _____

List all of the restraint systems found on this vehicle:

_____

_____

_____

PROCEDURE

1. Locate the air bag system in the service information. How are the critical parts of the system identified in the vehicle?

_____

_____

_____

_____

2. List the main components of the air bag system and describe their location.

_____

_____

_____

_____

_____

_____

_____

_____

3. Here are some very important guidelines to follow when working with and around air bag systems. These are listed next, with some key words left out. Read through these and fill in the blanks with the correct words.

   a. Wear _____ _____ when servicing an air bag system and when handling an air bag module.

   b. Wait at least _____ minutes after disconnecting the battery before beginning any service. The reserve _____ module is capable of storing enough energy to deploy the air bag for up to _____ minutes after battery voltage is lost.

   c. Always handle all _____ and other components with extreme care. Never strike or jar a sensor, especially when the battery is connected. Doing so can cause deployment of the air bag.

   d. Never carry an air bag module by its _____ or _____, and, when carrying it, always face the trimmed side of the module _____ from your body. When placing a module on a bench, always face the trimmed side of the module _____.

   e. Deployed air bags may have a powdery residue on them. _____ _____ is produced by the deployment reaction and is converted to _____ _____ when it comes in contact with the moisture in the atmosphere. Although it is unlikely that harmful chemicals will still be on the bag, it is wise to wear _____ _____ and _____ when handling a deployed air bag. Immediately wash your hands after handling a deployed air bag.

   f. A live air bag must be _____ before it is disposed. A deployed air bag should be disposed of in a manner consistent with the _____ and manufacturer's procedures.

   g. Never use a battery- or AC-powered _____, _____, or any other type of test equipment in the system unless the manufacturer specifically says to. Never probe with a _____ _____ for voltage.

Problems Encountered

_____

_____

_____

Instructor's Comments

_____

_____

_____

ELECTRICAL AND ELECTRONIC SYSTEMS JOB SHEET 50

Disarming and Enabling an Air Bag System

Name _____ Station _____ Date _____

NATEF Correlation

This Job Sheet addresses the following **MLR** task:

F.1. Disable and enable airbag system for vehicle service; verify indicator lamp operation.

This Job Sheet addresses the following **AST/MAST** task:

H.5. Disable and enable airbag system for vehicle service; verify indicator lamp operation.

Objective

Upon completion of this job sheet, you will be able to disarm and enable an air bag system so you can safely work around and near the air bag without accidentally deploying it and making it safe for the driver after you have serviced the vehicle.

Tools and Materials

Wiring diagram for the vehicle

Protective Clothing

Goggles or safety glasses with side shields

Describe the vehicle being worked on:

Year _____ Make _____ Model _____

VIN _____ Engine type and size _____

PROCEDURE

> **NOTE:** *This is a typical procedure. ALWAYS refer to the service information for the exact procedure. If you are following the directions given in the service information and they differ from the steps included in this worksheet, write the changes in procedure on the job sheet.*

1. Disconnect the negative battery cable.   Task Completed ☐

2. Tape the terminal of the cable to prevent accidental contact with the post   Task Completed ☐
   of the battery.

3. Remove the air bag system (SIR, SRS) fuse from the fuse block.   Task Completed ☐

4. Wait at least 10 seconds before proceeding.   Task Completed ☐

5. Disconnect the yellow connector at the base of the steering column.   Task Completed ☐

6. After service to the air bag or related system, reconnect the yellow con-   Task Completed ☐
   nector at the base of the steering column.

7. Install the fuse for the air bag system.   Task Completed ☐

8. Reconnect the negative battery cable.   Task Completed ☐

9. Perform a system self-test to make sure the system is enabled properly and ready for an emergency. How did you perform the self-test?

_____

_____

_____

Problems Encountered

_____

_____

_____

Instructor's Comments

_____

_____

_____

ELECTRICAL AND ELECTRONIC SYSTEMS JOB SHEET 51

Remove and Reinstall a Door Panel

Name _____ Station _____ Date _____

NATEF Correlation

This Job Sheet addresses the following **MLR** task:

F.2. Remove and reinstall door panel.

This Job Sheet addresses the following **AST/MAST** task:

H.6. Remove and reinstall door panel.

Objective

Upon completion of this job sheet, you will be able to remove and reinstall an inner door panel. This may be necessary to gain access to the various electrical and mechanical systems housed within the door.

Tools and Materials

Hand tools

Service information

Soft-faced hammer

Special tools for panel plug removal

Protective Clothing

Goggles or safety glasses with side shields

Describe the vehicle being worked on:

Year _____ Make _____ Model _____

VIN _____ Engine type and size _____

PROCEDURE

1. Install a power source so the electronic memories of the various accessories and computers can be maintained.   Task Completed ☐

2. Disconnect the negative cable of the battery.   Task Completed ☐

3. Locate any service precautions given in the service information, especially those pertaining to air bags and electronics. List and describe all precautions.

   _____

   _____

   _____

   _____

4. Which door panel are you planning on removing?

   _____

   _____

5. Again referring to the service information, summarize the procedure for removing the door panel you are about to remove.

   _____

   _____

   _____

   _____

6. Identify the components that will need to be removed in order to remove the door panel; include all speakers, switches, armrests, and compartments.

   _____

   _____

   _____

   _____

7. Begin your removal of the door panel by removing the screws and/or bolts that hold armrests and other major components to the panel. What did these retain?

   _____

   _____

8. Now examine all switches to determine if they should be removed once the panel is loose or prior to that. Describe your findings.

   _____

   _____

9. Remove all switches and everything else that may be directly attached to the panel.   Task Completed ☐

10. Starting at one end and using the correct tool, disengage the panel retaining clips from the door, being careful not to pull too hard or to damage the clips or the panel.   Task Completed ☐

11. Once the plugs are all free from their bores, lift the panel slightly out and up and check for anything that will interfere with the removal of the panel. What else needs to be removed?

   _____

   _____

12. Remove the panel and keep track of the retaining clips and plugs.   Task Completed ☐

13. Replace any plugs and clips that may have been damaged.   Task Completed ☐

14. Carefully position the panel so that the clips and plugs are over their respective bores, then press or lightly pound the panel with the clips into position.

Task Completed ☐

15. Reinstall all switches and everything else that was directly attached to the panel and was removed.

Task Completed ☐

16. Reinstall and tighten the screws and/or bolts that hold armrests and other major components to the panel.

Task Completed ☐

17. Reconnect the negative cable of the battery.

Task Completed ☐

18. Disconnect the electronic memory keeper.

Task Completed ☐

Problems Encountered

_____

_____

_____

Instructor's Comments

_____

_____

_____

ELECTRICAL AND ELECTRONIC SYSTEMS JOB SHEET 52

Using a Scan Tool on Body Control Systems

Name _____ Station _____ Date _____

NATEF Correlation

This Job Sheet addresses the following **AST/MAST** task:

H.7. Check for module communication errors (including CAN/BUS systems) using a scan tool.

Objective

Upon completion of this job sheet, you will be able to connect a scan tool to a body control system and retrieve trouble codes and check for communication errors.

Tools and Materials

Service information

Scan tool

Protective Clothing

Goggles or safety glasses with side shields

Describe the vehicle being worked on:

Year _____ Make _____ Model _____

VIN _____ Engine type and size _____

PROCEDURE

1. Briefly describe the vehicle network on your assigned vehicle.

   _____

   _____

   _____

   _____

2. Which network modules are on the high-speed network?

   _____

   _____

3. Which modules are on the low-speed network?

   _____

   _____

4. What happens to the vehicle network if the serial data line becomes grounded?

_____

_____

5. Where are the termination resistors located?

_____

_____

6. Does the system use a gateway module? _____ If it does, which module is it?

_____

_____

7. List the connections of the DLC that pertain to vehicle networking.

_____

_____

8. Using the scan tool, retrieve codes from the network and record them here.

_____

_____

9. For each code in the system, write a brief description of the problem detected.

_____

_____

10. Describe the diagnostic steps taken to find the problem. How are you going to verify the repair?

_____

_____

_____

_____

Problems Encountered

_____

_____

_____

Instructor's Comments

ELECTRICAL AND ELECTRONIC SYSTEMS JOB SHEET 53

Keyless Entry and Remote-Start System Operation

Name _____ Station _____ Date _____

NATEF Correlation

This Job Sheet addresses the following **MLR** task:

F.3. Describe the operation of keyless entry/remote-start systems.

This Job Sheet addresses the following **AST/MAST** task:

H.8. Describe the operation of keyless entry/remote-start systems.

Objective

Upon completion of this job sheet, you will be able to describe the operation of a vehicle's keyless entry and remote-start system.

Tools and Materials

Service information

Describe the vehicle being worked on:

Year _____ Make _____ Model _____

VIN _____ Engine type and size _____

PROCEDURE

NOTE: *There are many different designs of keyless entry and remote-start systems. Some are installed by the manufacturer, while others are aftermarket systems. The latter type is difficult to find information about. Check with the owner to see what documentation he or she may have.*

A. Keyless Entry Systems

1. Identify the type of entry system the vehicle has. Often it will be part of a key fob. Describe what is used on this vehicle.

   _____

   _____

2. What must be done to unlock and lock the doors from the outside?

   _____

   _____

3. List the actions that can be taken by the remote entry system.

_____

_____

4. Using the service information as a source, describe how the remote system works.

_____

_____

B. Remote-Start Systems

1. Describe how the system works, including the conditions that must be present to start the engine.

_____

_____

2. Describe how the system is activated.

_____

_____

3. Describe what happens if the remote system fails to operate.

_____

_____

Problems Encountered

_____

_____

_____

Instructor's Comments

_____

_____

_____

ELECTRICAL AND ELECTRONIC SYSTEMS JOB SHEET 54

Diagnosis of Instrument Panel Gauges, Warning/Indicator Lights and Maintenance Indicators

Name _____ Station _____ Date _____

NATEF Correlation

This Job Sheet addresses the following **MLR** task:

F.4. Verify operation of instrument panel gauges and warning/indicator lights; reset maintenance indicators.

This Job Sheet addresses the following **AST/MAST** task:

H.9. Verify operation of instrument panel gauges and warning/indicator lights; reset maintenance indicators.

Objective

Upon completion of this job sheet, you will be able to diagnose the operation of the instrument panel gauges, warning indicator lamps, and maintenance indicators.

Tools and Materials

Wiring diagram for the vehicle
Digital multimeter (DMM)
Scan tool
Service information

Protective Clothing

Goggles or safety glasses with side shields

Describe the vehicle being worked on:

Year _____ Make _____ Model _____

VIN _____ Engine type and size _____

PROCEDURE

1. Using the wiring diagram for the vehicle, describe how the instrument panel gauges operate.

   _____

   _____

   _____

2. Do the gauges receive instructions from a module? Do all the gauges operate in the same manner?

_____

_____

_____

3. Explain the operation of the parking brake warning lamp.

_____

_____

_____

4. Looking at the wiring diagram, is the brake warning "lamp" an actual bulb or is it a light-emitting diode?

_____

_____

_____

5. Explain the operation of the brake-warning lamp.

_____

_____

_____

6. Explain the operation of the "change oil" maintenance reminder.

_____

_____

_____

7. How is the "change oil" indicator reset?

_____

_____

_____

8. Connect the scan tool to the vehicle. Can any of the gauges or warning lamps be operated through the scan tool? List your findings.

_____

_____

_____

Problems Encountered

_____

_____

_____

Instructor's Comments

_____

_____

_____

ELECTRICAL AND ELECTRONIC SYSTEMS JOB SHEET 55

Verify Windshield Wiper and Washer Operation; Replace Wiper Blades

Name _____ Station _____ Date _____

NATEF Correlation

This Job Sheet addresses the following **MLR** task:

   F.5. Verify windshield wiper and washer operation; replace wiper blades.

This Job Sheet addresses the following **AST/MAST** task:

   H.10. Verify windshield wiper and washer operation; replace wiper blades.

Objective

Upon completion of this job sheet, you will be able to verify the operation of the windshield washer and wiper operation and replace wiper blades.

Tools and Materials

Replacement wiper blades or inserts

Service information

Protective Clothing

Goggles or safety glasses with side shields

Describe the vehicle being worked on:

Year _____ Make _____ Model _____

VIN _____ Engine type and size _____

PROCEDURE

Never operate the windshield wipers for an extended period on a dry windshield, otherwise wiper and windshield damage can occur.

1. Check the operation of the windshield washers. Do the wipers operate for a short period and then "park" as designed?

   _____

   _____

   _____

2. Try the wipers on intermittent. Do the wipers operate as designed?

   _____

   _____

   _____

3. Try low and high speed. Do the wipers operate as designed?

_____

_____

_____

_____

4. Using a wiring diagram, answer the following questions:

   a. If the wipers did not operate on low, and all other speeds operated normally, what is a possible cause?

   _____

   _____

   _____

   b. If the wipers fail to "park" after being shut down, what is a possible electrical/electronic cause?

   _____

   _____

   _____

5. Check the wiper blades for wear or damage. Do they streak the windshield?

_____

_____

_____

6. Explain the procedure to replace the wiper blades or inserts on the assigned vehicle.

_____

_____

_____

Problems Encountered

_____

_____

_____

Instructor's Comments

_____

_____

_____

ELECTRICAL AND ELECTRONIC SYSTEMS JOB SHEET 56

Identifying the Source of Static on a Radio

Name _____ Station _____ Date _____

NATEF Correlation

This Job Sheet addresses the following **MAST** task:

H.11. Diagnose radio static and weak, intermittent, or no radio reception; determine necessary action.

Objective

Upon completion of this job sheet, you will be able to correctly diagnose radio static and weak, intermittent, or no radio reception.

Tools and Materials

Jumper wire with alligator clips on both ends

Protective Clothing

Goggles or safety glasses with side shields

Describe the vehicle being worked on:

Year _____ Make _____ Model _____

VIN _____ Engine type and size _____

Describe the sound system in the vehicle:

_____

_____

_____

PROCEDURE

1. Turn on the radio and listen for the noise. Typically, the noise is best heard on the low AM stations. Describe the noise.

   _____

   _____

2. Listen to the radio with the engine off and with it running. Describe the difference in the noise level and the reception of the radio.

   _____

   _____

3. What can you conclude from your answer to number 2?

   _____

   _____

4. Operate the radio in AM and FM. Does the noise appear in both bands?

   _____

5. If the noise is only on FM, what could be the problem?

   _____

   _____

6. If the noise is heard on both AM and FM, continue the test by checking the antenna and its connections. Is the antenna firmly mounted and in good condition?

   _____

   _____

7. Check the connection of the antenna cable to the antenna. Are the contacts clean and is the cable connector in good condition?

   _____

   _____

8. Connect a jumper wire from the base of the antenna to a known good ground. Then, listen for the noise. Did the noise level change? Describe and explain the results.

   _____

   _____

9. Refer to the service manual and identify any noise suppression devices used on this vehicle. The noise suppression devices are:

   _____

   _____

10. Are all of the noise suppression devices present on the vehicle and are they mounted securely to a clean well-grounded surface?

   _____

   _____

11. Connect a jumper wire from a known good ground to the grounding tab on each capacitor-type noise suppressor. Listen to the radio and describe and explain the results of doing this:

   _____

   _____

12. Turn off the engine and disconnect the wiring harness from the voltage regulator to the generator. Start the engine and listen to the radio and explain the results of doing this.

_____

_____

13. Check the spark plug wires and spark plugs. Are both of these noise suppressor-types?

_____

_____

14. Check the routing, condition, and connecting points for the spark plug cables. Describe their condition.

_____

_____

15. Connect a jumper wire from a known good ground to the frame of a rear speaker. Listen to the radio and describe and explain the results of doing this.

_____

_____

16. What are your conclusions and recommendations based on this job sheet?

_____

_____

_____

_____

17. If the noise source has not been identified, what should you do next?

_____

_____

Problems Encountered

_____

_____

_____

Instructor's Comments

_____

_____

_____

ELECTRICAL AND ELECTRONIC SYSTEMS JOB SHEET 57

Diagnosing a Body Electronic System Circuit Using a Scan Tool

Name _____ Station _____ Date _____

NATEF Correlation

This Job Sheet addresses the following **MAST** task:

H.12. Diagnose (troubleshoot) body electronic system circuits using a scan tool; determine necessary action.

Objective

Upon completion of this job sheet, you will be able to troubleshoot a body control circuit using a scan tool.

Tools and Materials

Scan tool

Service information

Protective Clothing

Goggles or safety glasses with side shields

Describe the vehicle being worked on:

Year _____ Make _____ Model _____

VIN _____ Engine type and size _____

NOTE: *There is a wide variety of body control systems available on vehicles today, and depending on the vehicles available, not all of these parameters may be supported.*

PROCEDURE

Connect the scan tool to the vehicle and set-up according to the manufacturer's instructions.

1. Go to the scan tool BCM data list parameters and list the inputs that can be monitored with the scan tool.

   _____

   _____

   _____

   _____

   _____

   _____

2. If the scan tool is capable of displaying the ignition switch status, how could a technician use this information to help diagnose a problem with several devices that were not operating as intended?

_____

_____

_____

3. Can the door lock status be displayed? How could this information help a technician in diagnosing a problem with the power door locks?

_____

_____

_____

4. Is the scan tool capable of displaying the commands from the remote key fob? If the scan tool showed the proper command from the key fob and the power door lock switches, but the door locks would not operate, what might be a possible problem(s)?

_____

_____

_____

5. In #4 above, with the same situation, which components does the technician know are good, based on the scan tool info?

_____

_____

_____

BCM Output Control from the Scan Tool

1. List some of the BCM outputs that can be controlled by the scan tool.

_____

_____

_____

_____

_____

_____

2. Briefly explain how BCM outputs from a body electrical system can make it easier for a technician equipped with a scan tool to diagnose a problem.

_____

_____

_____

Functions that Have to Be "Learned" by the BCM

1. Using service information, list some of the body control functions that need to be reset or learned by the BCM, such as brake pedal position, window position, replacement keys or key fob remote controllers (others may be possible depending on the make and model of vehicle).

   _____

   _____

   _____

2. Choose one of the functions that have to be learned by the BCM and briefly describe the process of learning or calibrating a component or function into the BCM.

   _____

   _____

   _____

Difficulties Encountered

_____

_____

_____

Instructor's Comments

_____

_____

_____

ELECTRICAL AND ELECTRONIC SYSTEMS JOB SHEET 58

Diagnosing Anti-Theft Systems

Name _____ Station _____ Date _____

NATEF Correlation

This Job Sheet addresses the following **MAST** task:

H.13. Diagnose the cause(s) of false, intermittent, or no operation of anti-theft systems.

Objective

Upon completion of this job sheet, you will be able to diagnose anti-theft systems.

Tools and Materials

Digital multimeter (DMM)

Service information

Protective Clothing

Goggles or safety glasses with side shields

Describe the vehicle being worked on:

Year _____ Make _____ Model _____

VIN _____ Engine type and size _____

PROCEDURE

> **NOTE:** *There are many different designs of anti-theft systems, each with its own diagnostic and service procedures. Make sure you follow the guidelines given by the manufacturer before proceeding. This job sheet is designed to take you through the typical steps required to identify the cause of common problems.*

A. System Operation

1. Describe the theft system on the vehicle. Is the system meant to warn of a break-in, or is the system designed to shut the vehicle down so it cannot be driven?

   _____

   _____

2. Describe the major components of the system.

   _____

   _____

3. Describe the system inputs (such as door switches).

_____

_____

4. Describe the process to "arm" the system.

_____

_____

5. Describe the process to "disarm" the system.

_____

_____

B. System Diagnosis

1. State the customer's concern. Were you able to verify the customer's concern?

_____

_____

2. Search for any relevant bulletins on the customer's concern. Describe what you found.

_____

_____

3. Using the scan tool, check for any DTCs relating to the concern. Summarize what you found.

_____

_____

4. While the scan tool is connected, check to ensure that there are no codes for module communications. Describe your findings.

_____

_____

5. If a code was stored that relates to the theft deterrent problem, summarize your diagnosis here.

_____

_____

6. If there are no codes stored, go to the symptom diagnosis charts. Were you able to determine the cause of the problem? If so, summarize your findings here.

_____

_____

Problems Encountered

_____

_____

_____

Instructor's Comments

_____

_____

_____

ELECTRICAL AND ELECTRONIC SYSTEMS JOB SHEET 59

Reflashing an Electronic Module

Name _____ Station _____ Date _____

NATEF Correlation

This Job Sheet addresses the following **MAST** task:

H.14. Describe the process for software transfers, software updates, or flash reprogramming on electronic modules.

Objective

Upon completion of this job sheet, you will be able to update the software in a vehicle's electronic control module.

Tools and Materials

Scan tool

PC with Internet capabilities

Service information

Protective Clothing

Goggles or safety glasses with side shields

Describe the vehicle being worked on:

Year _____ Make _____ Model _____

VIN _____ Engine type and size _____

PROCEDURE

1. Connect the scan tool to the vehicle's DLC and retrieve the BCM's part number or the vehicle's VIN and record them here.

   _____

   _____

2. Disconnect the scan tool from the vehicle and connect it to a PC that can link the scan tool to the flash software. Some scan tools will connect directly to an Internet site. What is your source for the updated software?

   _____

   _____

3. Enter the part number for the BCM in the appropriate field in the PC and select "Show Updates" on the menu.

_____

_____

4. Select the desired flash line. Press the Next button to begin downloading the software. What flash line did you select?

_____

_____

5. Monitor the progress of the downloading to the scan tool. Once downloading is complete, connect a special battery charger with a clean voltage output or a "backup" battery to the vehicle's battery to insure about 13.5 volts at the battery. Why should you do this?

_____

_____

6. Connect the scan tool to the DLC and turn the scan tool on. Move through the menus on the scan tool until the desired flash screen is shown. Then, follow all instructions given on the tool. What are those instructions?

_____

_____

_____

Problems Encountered

_____

_____

_____

Instructor's Comments

_____

_____

_____

— NOTES —

— NOTES —